At Issue

The Energy Crisis

Other Books in the At Issue Series:

AIDS in Developing Countries

Age of Consent

Are Natural Disasters Increasing?

Does Outsourcing Harm America?

Does the Two Party System Still Work?

The Federal Budget Deficit

How Does Religion Influence Politics?

Is Iran a Threat to Global Security?

Nuclear and Toxic Waste

Piracy On The High Seas

Should the Federal Government Bail Out Private Industry?

Should the Internet Be Free?

Should the U.S. Close Its Borders?

Should Vaccinations Be Mandatory?

Voter Fraud

At Issue

The Energy Crisis

Lisa Krueger, Book Editor

GREENHAVEN PRESS
A part of Gale, Cengage Learning

GALE
CENGAGE Learning™

Detroit • New York • San Francisco • New Haven, Conn • Waterville, Maine • London

AUG 1 1 2010

Christine Nasso, *Publisher*
Elizabeth Des Chenes, *Managing Editor*

For more information, contact:
Greenhaven Press
27500 Drake Rd.
Farmington Hills, MI 48331-3535
Or you can visit our Internet site at gale.cengage.com

Articles in Greenhaven Press anthologies are often edited for length to meet page requirements. In addition, original titles of these works are changed to clearly present the main thesis and to explicitly indicate the author's opinion. Every effort is made to ensure that Greenhaven Press accurately reflects the original intent of the authors. Every effort has been made to trace the owners of copyrighted material.

Cover image copyright © Debra Hughes 2007. Used under license from Shutterstock.com.

LIBRARY OF CONGRESS CATALOGING-IN-PUBLICATION DATA

The energy crisis / Lisa Krueger, book editor.
        p. cm. -- (At issue)
    Includes bibliographical references and index.
    ISBN 978-0-7377-4679-2 (hardcover) -- ISBN 978-0-7377-4680-8 (pbk.)
    1. Power resources--Juvenile literature. 2. Energy policy--Juvenile literature.
    I. Frohnapfel-Krueger, Lisa.
    HD9502.A2E54359 2010
    333.79--dc22
                                                            2009047745

Printed in the United States of America
1 2 3 4 5 6 7 14 13 12 11 10

Contents

Introduction   7

1. By 2030 the Demand for Oil Will   9
   Exceed Available Supply
   Paul Beingessner

2. There Would Be No Energy Crisis If   13
   the World Consumed Energy Efficiently
   Michelle Moore

3. Oil Can Be Replaced by Clean Energy   17
   William Marsden

4. Investment in the Oil Industry Is Needed   26
   to Meet Future Energy Demands
   Syed Rashid Husain

5. Nuclear Power Is a Viable Solution   31
   to the Energy Crisis
   Jeffery D. Kooistra

6. The Looming Energy Crisis Creates   38
   Opportunities for Solar Power
   James M. Higgins

7. Global Food Prices Have Increased Due   49
   to the Energy Crisis
   Jacqui Fatka

8. Fear Blocks Opportunities to Resolve   54
   the Energy Crisis
   James Lovelock

9. The Peaking of Oil Is Causing Tension   60
   Between Nations
   Tony Black

10. Industrialized Nations' Economies Could   68
    Suffer If the Oil Supply Is Blocked
    Gulf Daily News

11. Natural Disasters and Oil Supply Failure   76
    Could Cause a Global Crisis
    The Economist

12. Modernizing Electricity Management Is   81
    Key to Meeting Energy Demands
    Galvin Electricity Initiative

13. IT Companies Become Energy Efficient   85
    in face of Global Energy Crisis
    T. E. Raja Simhan

Organizations to Contact   91
Bibliography   96
Index   101

Introduction

The demand for energy is growing as the world's population increases and developing economies expand. Current sources of energy include fossil fuels; renewable sources of energy such as wind, solar, and water power; and nuclear power. All energy sources have both advantages and disadvantages.

Fossil fuels are sources of energy that are generated from natural resources such as oil, gas, and coal. Fossil fuels were formed millions of years ago under the earth's surface. Today, fossil fuels are extracted from the earth to generate energy such as electricity, fuel for transportation and manufacturing, and to make pesticides and commercial fertilizers. Society's dependence on fossil fuels cannot be overstated. The majority of the world's electricity is generated from coal, and many products used in contemporary society depend on oil and natural gas. Fossil fuels are widely used because they are cost effective and functional. When fossil fuels are converted to energy, however, pollutants are released into the environment, raising concerns of air and water pollution as well as global warming. Unlike renewable energy sources, the supply of fossil fuels is not infinite. As the world's demand for fossil fuels increases, their availability is expected to dwindle.

Renewable energy is generated from natural resources such as sunlight, wind, tides and geothermal heat. Extracting energy from these sources is favorable because their supply is infinite and they pose a minimal threat to the environment. Generally, renewable energy emits little to no air pollution during operation. The greatest impact on the environment from renewable energy is how the components to power renewable energy are manufactured, installed, and disposed of. For instance, the manufacturing and installation of solar panels requires the use of fossil fuels. The disadvantages of renewable energy include land use conflicts, efficiency, and cost. Further development is needed to make renewable energy more affordable and practical.

Nuclear power plants for generating electricity are located throughout the world. Uranium, a radioactive material found in rocks, is used to power nuclear power plants. Nuclear power plants are cost effective and do not pollute the air with nitrogen oxides, sulfur oxides, or dust, nor do they emit greenhouse gases. Nuclear accidents result in radioactive waste that may be dangerous for thousands of years, however. The most severe radioactive accident in the world occurred in 1986 at Chernobyl, in the former USSR. According to the World Nuclear Association, "The Chernobyl accident in 1986 was the result of a flawed reactor design that was operated with inadequately trained personnel and without proper regard for safety." Nuclear power plants have numerous safety measures and emergency systems in place to prevent accidents. Some experts believe nuclear energy has the potential to drastically reduce the world's dependence on fossil fuels.

Government and public support is needed to develop new processes to harness the energy of fossil fuels, renewable energy, and nuclear power. Many people believe the world is already experiencing an energy crisis due to the negative impact current energy sources have on the environment and the dependency the world has on countries with an abundance of fossil fuels. Others believe the world is not experiencing an energy crisis, and they argue that people need to use current sources of energy more efficiently. Scientists and researchers are actively seeking methods and resources that will make the energy supply not only healthy and safe but also meet the world's increasing demand for energy.

1

By 2030 the Demand for Oil Will Exceed Available Supply

Paul Beingessner

Paul Beingessner was a transportation consultant and weekly contributor to the Briar Patch *magazine. His weekly column focused on farming and transportation issues affecting farmers and communities throughout the world.*

Denying that oil consumption will exceed supply is destructive. By 2030 the world will experience a severe oil shortage. Politicians and individuals in positions of power must address the pending oil shortage instead of protecting people profiting from the current energy system.

In Freudian psychology, denial is an ego defence mechanism. Defence mechanisms are used by our subconscious minds to help us deal with realities we find too difficult to face. They aren't signs of mental illness; everyone uses them to some extent. Using them . . . becomes a problem when it leads us to do things that are harmful in the end. Freud had a whole list of defence mechanisms, ranging from repression to projection to denial. The key thing to remember about defence mechanisms is that they operate in the subconscious. We don't realize we are using them. So denial is not really a form of lying to ourselves; we actually believe what we are saying.

Now, Freud is kind of old hat these days. Many psychologists today figure he was too influenced by the repressed

Paul Beingessner, "Freud and the Energy Crisis," *Briar Patch*, vol. 37, September/October 2008, p. 32. Reproduced by permission.

upper-middle-class women he saw in his therapy sessions, so they dismiss most of his theories as unscientific. In the case of denial, however, current human behaviour is making a pretty good case for Freud's idea.

The Fossil Fuel Reality

How does all this psychological mumbo jumbo relate to everyday activities like agriculture? Agriculture in developed countries, and increasingly in other countries, is rooted in the consumption of fossil fuels. Fossil fuels are the basis of planting, weeding and harvesting operations. They produce our fertilizers and pesticides[, and they] transport our crops to processors and markets. And they are turning out.

Yes, you can trot out all the figures you want about oil sands, shale oil, coalbed methane and the like, but a friend recently put it all in perspective for me when he told me the following: currently, the world uses 87 million barrels of oil per day. This is increasing rapidly and, if supplies manage to keep pace with demand, is projected to reach 116 million barrels by 2030. At that time, we will be using a trillion barrels every 23 years. In all of history up until now, we've used a trillion barrels. By 2030, we'll be using this much every 23 years. It's clear that our bingeing on fossil fuels will come to an end a lot sooner than we may think.

Fossil fuels are the basis of planting, weeding and harvesting operations. They produce our fertilizers and pesticides[, and they] transport our crops to processors and markets. And they are running out.

Whenever this is pointed out, though, the usual response by those in the throes of Freudian denial is to point to the reserves of oil sands in Alberta and parts of Saskatchewan. Total reserves in the Athabasca formation are estimated at around two trillion (2,000 billion) barrels, but only about 170 billion are recoverable with current technology. Even if the total amount were recoverable (which it won't be) and notwithstanding the massive amount of natural gas or other forms of energy required to extract the oil, it is plain that we will indeed run out of oil.

Of course, long before that happens, the price will skyrocket and $1.40 per litre diesel fuel will be a distant, cherished memory.

The figures above come from the International Energy Agency (IEA). The IEA grew out of the Organization for Economic Co-operation and Development and is controlled by 27 developed countries including Canada, the U.S., Japan and the U.K. The IEA's figures are accepted by and available to these countries. So why do most of them still act as if the petroleum age will go on forever?

The Uncertain Future

The end of low-cost oil will mean changes to our world as dramatic as those brought on by the advent of low-cost oil. If we hope to maintain anything resembling our current level of material comfort we will continue to need large amounts of energy, but we will have to get them somewhere else. Some folks point to nuclear power as a partial solution to our impending shortages. But guess what? There are only 60 years of known uranium reserves for the reactors currently operating in the world.

The simple fact, almost completely neglected by politicians and the public in general, is that we need to cut energy consumption, dramatically and rapidly. Doing so will stretch out existing petroleum reserves and allow us more time to change the way we do just about everything, including producing food.

And here is where denial really comes into its own. Reducing energy consumption will mean fewer and smaller vehicles, less miles driven, fewer airplane trips, fewer exotic vacations, fewer leaf blowers, less food imported by air from Taiwan and Chile, smaller houses, and no more electric toothbrushes. It will mean a whole lot more than that as well. But these are precisely the things we don't seem prepared to give up. In fact, we in North America, 1.3 billion Chinese, and 1.1 billion Indians seem to think these things are measures of the good life. Believing this, we find all kinds of ways to tell ourselves that the impending end of oil won't happen, or that there will be a technological solution that will allow the orgy of energy use to continue.

The end of low-cost oil will mean changes to our world as dramatic as those brought on by the advent of low-cost oil.

I said earlier that the failure to acknowledge and act on these realities is a form of denial by our politicians and ourselves. In fact, that statement is denial in itself. The truth is that politicians, at least those in real positions of power, know all this. What they are about is protecting the interests of those reaping massive profits from energy production in the current system.

Our denial, as citizens, is real though. In Freud's world, denial was ... pathological [only] if it caused you to act in self-destructive ways. Seems to me we passed that point some time ago.

2

There Would Be No Energy Crisis If the World Consumed Energy Efficiently

Michelle Moore

Michelle Moore is senior vice president of the U.S. Green Building Council in Washington. The Christian Science Monitor *was founded in 1908 to provide fair, balanced news in response to the "yellow" journalism of the day, and it has become an international online and print news organization.*

Using energy efficiently and using renewable energy sources can provide real solutions to the energy crisis. This has been proven in the area of buildings, which can be made so energy efficient that they save more money than the cost of implementing the energy-saving measures. Abundant energy is avialable from the renewable sources such as the sun and wind. Making efficient use of these renewable resources can create an energy infrastructure that is able to perform the same critical fucntions as the current centralized power plant infrastructure, but with more redundancy and resilience.

Washington—To drill or not to drill is the wrong question.

Real solutions to the energy and climate crisis are available today if we focus on what we have in abundance instead of arguing over what's exhaustible and dwindling—namely fossil fuels.

The Christian Science Monitor, October 6, 2008 for "To Drill or Not to Drill Is Not the Question," by Michelle Moore. Copyright © 2008. Reproduced by permission of the U.S. Green Building Council. www.usgbc.org.

Emphasizing efficiency and renewables, in that order, is already working in the building sector, which is demonstrating how we can change for the better by changing the way we think.

There are plenty of opportunities to meet energy demand through efficiency.

Buildings represent almost 40 percent of US energy consumption and an equivalent percentage of CO_2 emissions—more than automobiles or industry.

Research shows that the average additional cost for new LEED-certified buildings is less than 2 percent of total project costs, and that those costs are repaid within the first 12 months of occupancy through operational savings.

We have the ability to offset 85 percent of America's incremental electricity needs in 2030 through building and appliance efficiency measures that save more money than they cost to implement, as McKinsey & Co. articulated in "Reducing US Greenhouse Gas Emissions: How Much at What Cost?"

These solutions—simple things, such as retrofitting commercial buildings, weatherizing our homes, and using energy-efficient appliances—challenge every American and every American business to act. The immediate reward: We improve our bottom line. Already, more than 24,000 homes, schools, and offices are registered with the US Green Building Council's LEED rating system.

Research shows that the average additional cost for new LEED-certified buildings is less than 2 percent of total project costs, and that those costs are repaid within the first 12 months of occupancy through operational savings. For example, Adobe Systems spent $1.4 million greening three LEED Platinum office towers at their headquarters in San Jose, Calif. The investment was fully paid back in just 10 months through energy, water, and other operating efficiencies.

We know renewable energy is limitless. But to put it in stark terms; the earth gets enough solar power every 40 minutes to meet the whole world's energy demand for a full year. Wind power and geothermal resources are similarly capable of providing for our needs.

What's more, as demand for renewables rises, costs come down. That's the exact opposite of oil, which gets more and more expensive as demand goes up. While the cost of oil rose to more than $100 per barrel, the cost of silicon used to make solar cells dropped more than 80 percent.

By taking what may seem like a radical step now, and making our homes, schools, and offices power producers instead of energy hogs, people are demonstrating how effective putting these facts into practice can be. By making efficiency and renewables the norm, we can envision a very different future.

But to put it in stark terms; the earth gets enough solar power every 40 minutes to meet the whole world's energy demand for a full year.

Buildings can become part of a distributed, renewable energy infrastructure that's not only cleaner and greener, but also more resilient than one that depends on a small number of big power plants to keep the lights on.

Highly redundant distributed systems like these—systems with a network of interconnections between many parts capable of performing the same critical functions—are not only common in nature; they are increasingly preferred in commerce. The triumph of the Internet over mainframes and wireless over wires speak to that fact.

Successful steps forward are not yet ubiquitous, but they do exist from Maryland to Seattle. Early leaders like Austin,

Texas and the State of California focused first on energy efficiency, demonstrated the case through government-owned buildings and private-sector incentives, built a constituency for change, and are now pushing the frontier with zero-energy buildings.

The city of Austin, which boasts the first green home program in the country, now requires all new homes to be net-zero energy capable, or able to operate completely off the grid, by 2015; and California's Energy Commission has recommended similar requirements by 2020.

Realizing the full potential of efficiency puts dollars back into our economy. Those can then be invested in renewable energy technologies to build scale and bring down costs. We have what we need. It is not necessary to argue over drilling.

By embracing efficiency and deploying renewables right now, we can help build a prosperous and sustainable low-carbon future.

3

Oil Can Be Replaced by Clean Energy

William Marsden

William Marsden is an award-winning senior investigative reporter for the Gazette *in Montreal.*

Politicians and those in political power have the ability to remedy the world's energy crisis. With the right government policies, clean energy can become a reality. Germany's success with wind energy is an example of how government policies can facilitate the use of clean energy.

To say that this week's [January 2009] International Auto Show in Detroit marked a new beginning might be going too far, given the history of the industry. But at the very least it was a sign. A new direction taken. A light at the end of the tunnel. The carmakers' fresh lineup of electric and hybrid vehicles, and plenty of shiny promises of more to come, offered a glimmer of hope to a world stuck on oil. The message was simple. The future of clean energy is now. Viable alternatives to the combustion engine exist. We have liftoff.

Resistance to Change

Two questions: Why did it take so long? And if this is so, do we really have an energy crisis or is the problem elsewhere? With concerns over global warming and pending oil shortages growing, not to mention intense air, ground and water pollution on a local and international level, events such as the auto show have put a lie to the doomsday scenario that nothing can replace oil.

In other words, we have the technology to move toward a life without oil. What we don't appear to have is the political will.

Which is odd given that petroleum experts are warning of coming oil shortages—some say they are already here—as the world's reserves are fast depleted by increased consumption particularly from China and India. At the same time, there is increasing urgency to reduce as soon as possible by at least 80 per cent our greenhouse gas emissions.

We have the technology to move toward a life without oil. What we don't appear to have is the political will.

The reason behind this reluctance to change, many experts contend, is that politicians and policy-makers are still stuck in the old paradigms shaped by the dictates of using fossil fuels for energy. In other words they are tied to an energy source that demands the huge capital expenditures to extract and refine it, plus an enormous global distribution system to get it to market. All of which means an energy world led by big business and power elites.

What if you had an energy source that empowered the little guy? Experts say that that's essence of clean energy. It's like the dawn of the Internet. It presents a whole new paradigm that literally gives power to the people.

Making Clean Energy a Priority

Just ask men like Dr. Hermann Scheer, the German politician whose "sheer determination" is transforming his country into a clean energy Goliath, creating along the way a $13-billion alternative energy industry driven by wind, solar and biomass energy that employs more than 200,000 people.

In speeches around the world, Scheer has been driving home the point that Germany's ongoing transformation is being achieved not through superior technology or know-how, but by enacting the right government policies, by pushing aside the old ways and bringing in systems geared specifically to clean energy.

Germany's new regulations, which the country enacted in 2000, helped create a whole new economic framework that has given clean energy a priority role in the country's power generation. Hence the name: the Law for the Priority of Renewable Energy.

Germany's ongoing transformation is being achieved not through superior technology or know-how, but by enacting the right government policies.

The law has three elements. First, the law guarantees that clean energy producers have priority when it comes to selling to the power grid.

"No grid company can say, 'No, we don't take it, we have other contracts,'" Scheer said. "It has absolute priority. It must be taken first." Second, the power grid owners must pay a fee for the clean energy that guarantees a profit to the producer. "If you have too low a fee, then there will be no investments," Scheer said.

Finally, the law guarantees unlimited access to the grid no matter how much or how little clean energy you supply. This means that even the smallest one-kilowatt supplier has priority access to the grid.

The law essentially created autonomy for clean energy producers. Not only did they no longer have to beg the big power distributors for access to their grids, but they were also assured a profitable tariff.

Were identical laws enacted in Quebec, the scenario would go something like this: You could install solar panels on your home, hook them up to the Hydro-Quebec grid and Hydro-Quebec would have to pay you a profitable tariff for any surplus energy you feed it.

For Germany, the result is the country has increased its clean energy to 18 per cent of total electrical capacity, from 4.5 per cent.

Making the Case for Wind Power

"We have had 3,000 megawatts of renewable energy installation coming onstream every year since 2001," Scheer said in an interview. "In seven years, 22,000 megawatts (MW) altogether. And we created a new industry. It has opened the power grid to everybody. Fifty per cent of the owners of windmills are small farmers." Germany's windmills produce an average of 1.2 MW per mill. "The newest windmills are 6 MW," he said. "That means that if we were to replace these 20,000 windmills by windmills in the average of 6 MW, we could have seven times more production. That means practically 50 per cent of (Germany's) power consumption.

"This shows that (clean energy) can be introduced much faster than most experts think." To put that another way, Quebec plans to spend $7 billion on a project to build four dams on the province's last remaining free-flowing river, the Romaine River on the North Shore. The project will create an additional 1,500 MW and will take up to 10 years to bring on line. Germany creates that much power in clean energy every six months and doesn't destroy a river in the process.

What's more, wind power costs about half as much as power from a dam, and a large windmill can be up and running in one week. A megawatt will produce enough electricity to power about 300 homes for a year.

According to the Canadian Wind Energy Association [CWEA], Quebec's upper North Shore is an excellent location for wind power with some of the highest and most reliable average wind velocities in the country.

In fact, this is true for most regions of Canada.

"We have a tremendous wind resource that is one of the top two in the world," CWEA president Robert Hornung said.

Critics claim that wind energy is problematic because winds are unreliable. What do you do when the wind dies? In Canada, there is an easy solution. We get about 60 per cent of our electrical power from hydro. In Quebec the figure is 98 per cent. Hydro power is an ideal backup for wind power because it can decrease or increase its generation in seconds. Nuclear or coal-fired power plants always have to be working.

The key is to assure that peak demands are met. Wind farms can handle this situation as long as they are wisely distributed around the province. There is always wind blowing somewhere.

"(Clean energy) can be introduced much faster than most experts think."

"With the combination of hydro power and wind energy," Scheer said. "Canada could within 10 years have a complete, renewable, clean energy power system." Wind and hydro nicely complement each other, Hornung said. In winter when river flows fall drastically, wind power is at its strongest.

"We have a huge hydro-electric grid that is very helpful for facilitating the integration of wind," Hornung said. "We sit next door to the world's biggest electricity market, which provides significant export opportunities. And we have a manufacturing sector that's struggling and is looking for opportunities to get involved in the technologies of the next century. So we think there is a great opportunity for Canada but it requires us to think big." Normand Mousseau, a physics professor at the Universite de Montreal who researches alternative energies, said that to maximize clean energy North American electricity grids should be integrated. If there was a shortage of wind power in one region, the slack could be picked up by another.

The CWEA last year set a goal for Canada of 20-per-cent wind energy by 2025 and is lobbying governments to enact the same kind of enabling legislation that Germany did.

Wind power now supplies slightly more than one per cent of Canada's electric power with 85 wind farms creating 2,369 MW of generating capacity—enough power to meet the needs of 674,000 homes.

Hornung admits the 20-per-cent target is modest to the extreme. Denmark has already reached that level and Spain is well on its way. Germany has seven per cent wind power and 18 per cent overall clean energy power. Quebec produces 531 MW from wind, which is less than one per cent of its total capacity of 40,000 MW. The utility wants to increase wind capacity to 10 per cent of its total grid.

Mousseau is convinced that Canada could easily get almost all of its energy from wind power. It's just a question of political will.

"The technology is there," he said. "There is enough wind power, in fact there is plenty of wind power in Quebec alone to power most of North America." He said the problem is that the people who run Hydro-Quebec either are geared to building dams and don't regard wind power as "real electricity." Scheer knows this story well. "We adopted (the renewable energy law) against a lot of opposition from the power companies." Canada big power elite are primarily public utilities and oil and gas companies. They control the grids, making it difficult to establish independent and autonomous clean energy companies. Most of our green energy comes from the practice of public utilities parceling out a small number of clean energy contracts. And where do they go? Mostly to foreign companies from Germany, Spain and Denmark, where clean energy thrives.

These countries also get large energy contributions from solar, geothermal and biomass. In Germany, solar supplies 5,000 MW, which is equivalent to more than three Romaine River hydro projects.

Looming oil shortages and global warming mean it is critical to act quickly, Mousseau said.

"There could be a huge upheaval if people don't move fast," he said. "We have to do it now before oil gets back to high prices . . . If we don't act, then we will be in deep trouble because a lot of money will be used to buy oil instead of building new technology infrastructure and moving in the right directions." Which leads us back to cars and the air, ground and sea transportation system that relies almost exclusively on fossil fuels, without which our global economy would shut down.

Life After Oil

What is available that will be viable alternatives to oil? As the Detroit auto show demonstrates, the answer is quite a bit.

Hybrid and electric plug-in vehicles are the way of the future and the faster we switch over to them the better, says Sheldon Williamson, an engineering professor and vehicle battery expert at Concordia University.

He says that batteries and ultra-capacitors using nanotechnologies [machines and materials made on a submicroscopic scale] will soon get us to Toronto on one charge.

Electric buses run by capacitors are viable alternatives to our diesel buses, he said.

"I could load the bus up with all ultra-capacitors," he said. "No batteries. Forget about batteries. For a bus that makes short stops, ultra-capacitors are enough." The bus could go at least five stops, about a half-kilometre apart, then recharge in less than a minute at a quick-charging station at a bus stop as passengers get on and off.

Hybrid cars already promise 50 kilometres to a litre of gas. "For most trips you will be able to run on only electricity," Mousseau said.

Williamson said there is plenty of nickel and lithium in the world to run car batteries for years to come. Car manufacturers such as Toyota and Honda use nickel-metal hydride batteries, which are recyclable.

Public transit, intercity trains, cars and trucks can all be powered by electricity or hybrid systems.

But that's not the case with air and sea transport. At the moment, sailboats, blimps and gliders are the only answers, which are no answers at all. We'll probably just have to become more local.

The overall gripe about alternative energy sources is that they are expensive. Anybody who has shopped around for a hybrid knows that they cost several thousand dollars more than a gas guzzler.

All clean energy is in the long run cheaper because you don't have to pay for the fuel.

But that is temporary. New technology is always more expensive until mass production takes over.

This is what has already happened with wind power, which is cheaper than any other power source, Scheer said.

"All clean energy is in the long run cheaper because you don't have to pay for the fuel," he said. "Renewables are the only thing that can lead to clean and cheap energy forever." So is there really an energy crisis, or is it simply a crisis of ignorance? Human potential is the most overlooked and most important factor in overcoming the problems of peak oil and climate change, Scheer said. "We cannot discuss this question only on cost comparisons. Who has the courage in 20 or 25 years to tell their own children, 'We could have solved the problem but it was too expensive for us. The additional cost of three or four cents per kilowatt/hour was too costly for us.' That's really shabby."

4

Investment in the Oil Industry Is Needed to Meet Future Energy Demands

Syed Rashid Husain

Syed Rashid Husain is Vice President at Al-Azzaz, a construction and trading firm in Saudia Arabia.

Global oil prices are falling due to the economic downturn. As oil prices fall, the incentive to invest in the oil industry declines. Investment dollars are used to produce, refine, and transport oil more efficiently. The oil industry needs investment to maintain the supply of oil.

Oil markets are facing a major slump—for a number of reasons—and continue to stream further down. As I write these lines, prices are already in the vicinity of $50 a barrel. Rather than seeking a ceiling, crude markets now appear looking for a floor—somewhere—at respectable levels. What a transition indeed. And indeed this transformation is not without ramifications, of considerable magnitude, one can easily deduce.

Crude markets have entered a phase where, due to low prices, the incentive to invest in the industry is getting less and less.

And if the trend continues, as some are arguing today, another round of price spiral may not be far off. The emerging scenario may not only be disastrous for the industry, but indeed for the overall global energy balance too—a real cause of concern indeed. We need to wake up to the consequences now—and not later.

Falling Prices

Global crude prices have fallen by two thirds over the past four months, and it appears set to fall further—unless drastic steps are in place. Interestingly, it took 40 months for oil prices to rise from $50 a barrel to almost $150 a barrel and just four months for them to crash to the current lows.

The London-based Center for Global Energy Studies (CGES) now believes that the global oil demand would contract in 2008 for the first time in a quarter of a century.

In order to keep wheeling this crude-driven civilization of ours, investment in the industry is a must.

The dwindling fortunes of oil may indeed have brought smiles in some major global capitals. Many must be heaving a sigh of relief at the turn of events.

Yet, these are not the best of times for the industry. Energy is indeed a costly affair. In order to keep wheeling this crude-driven civilization of ours, investment in the industry is a must. And with falling prices, this investment is now in question. Would there be enough investments to keep meeting the growing global requirements, is a billion dollar question.

Early warning shots are already there—investments in the industry are getting shy. With the global energy requirements continuing to grow—albeit at a slower pace than before—the issue of meeting the future global needs is a real one. From where [would] the incremental supplies . . . come, in case investments in the sector continue to be bogged down—due to the lack of incentives?

Platts, the energy information provider, had also projected last year that companies that produce, refine and transport oil and natural gas will need as much as $21.4 trillion in capital expenditures through 2030 to meet the world's growing energy demands.

In its recently released World Energy Outlook, the OECD [Organization for Economic Cooperation and Development] energy watchdog IEA [International Energy Agency] underlines that more than a trillion dollars in annual investments is needed to find new fossil fuels over the next two decades to avoid the impending energy crisis that could easily choke the global economy.

At a time when major oil companies are pulling back investments in view of one of the most severe economic downturns in a generation, and lack of incentive to invest in the sector, the IEA stressed that it's vital for continued investment in new projects. The total potential tab through 2030 as per the IEA is [on] the order of $26.3 trillion—colossal by any means.

"There remains a real risk that underinvestment will cause an oil supply crunch" by 2015 as the decline in output from mature oil fields speeds up.

Consequences of Underinvestment

"There remains a real risk that underinvestment will cause an oil supply crunch" by 2015 as the decline in output from mature oil fields speeds up, the Paris-based adviser to 28 oil-consuming nations said in its annual report. "The current financial crisis is not expected to affect long-term investment, but could lead to delays in bringing current projects to completion."

OPEC [Organization of the Petroleum Exporting Countries] has also warned that crucial downstream investment—in refining and distribution—will be curtailed if the oil price is not maintained at a reasonable level.

The crisis is beginning to unfold. The prevailing uncertainty is already prompting companies to withhold billions of dollars of investment in new oil field and refining projects. Royal Dutch Shell PLC, Europe's largest oil company, said last month it was pushing back a decision on expanding an oil sands project in Canada. North American refining giant Valero Energy Corp. has also announced curtailing capital spending for the rest of 2008 and 2009. Also, Marathon Oil Co. said it has delayed expansion of a gasoline refinery in Detroit "due to current market conditions."

Saudi Aramco has also fired the early warning shots. It's Chief Executive Officer Abdullah Jum'ah in a handout distributed earlier this month at an industry summit in Beijing said that a further drop in crude prices may curtail investments needed to offset declining output in aging fields. Investment is also needed to expand production capacity to meet long-term demand growth, Jum'ah emphasized in his message. And prices have dropped further since the early November salvo by Jum'ah.

Saudi Aramco was reportedly already "reviewing" parts of its $129 billion upstream push to boost oil production in light of the "current economic circumstances," recent press reports said. Khaled Al-Buraik, an executive director at Saudi Aramco, was recently quoted as saying that though Aramco's short-term projects were on track and the Kingdom [of Saudi Arabia] would reach its target of increasing production capacity to 12.5 million barrels a day by the end of next year, but the development of the Manifa field, which was to add 900,000 barrels per day of capacity by 2011, was under review.

Things are definitely under evaluation. The comments indicate that sustained lower oil prices would not only impact the high cost [of] non-OPEC or the Sand tar projects but could also affect future production even in an area of the world where oil production is comparatively cheap and easy.

"It is clear that collapsing oil prices are not only detrimental to the economies of oil-producing states, but also to future upstream investments to sustain future oil demand consumption," Vienna-based consultant JBC Energy said in a recent market report.

Demand for oil and crude prices may be falling with the economic slowdown, but that could well lead to a supply-side crunch in the next year or so, and that will push oil prices higher again.

And that is the big challenge. The industry needs to be prepared for tomorrow, even in these uncertain times. If we do not act now, another round of price spiral may not be far off. Is anyone taking cognizance?

5

Nuclear Power Is a Viable Solution to the Energy Crisis

Jeffery D. Kooistra

Jeffery D. Kooistra is one of the two authors of "The Alternate View" column in Analog Science Fiction & Fact. *He is a former associate editor of* Infinite Energy Magazine *and continues to stay active in the study of alternative physics and the paranormal. He resides in Grand Rapids, Michigan, with his wife and three children.*

Over the past thirty years, politicians and energy companies have failed to adequately address long-term energy needs. Fossil fuels and renewable energy such as solar and wind energy are not the answer to the world's energy needs. Nuclear power is a viable solution because it is cleaner than fossil fuels and the technology to generate nuclear energy is already in place.

I got so angry when I drove into the station and saw the price on the pump. I needed gas so I couldn't just drive away, but this was ridiculous! Gasoline should not cost this much! As I was pumping fuel, I groused about the cost some more, and wondered if I should fill up or hope the price would go down in a few days and only fill the tank half way. I was very frustrated with "our leaders"—those people in business who run energy companies, and in government those whom we elect to anticipate and solve problems before they occur.

Jeffery D. Kooistra, "Energy Crisis Redux: A Polemic," *Analog Science Fiction & Fact*, vol. 129, January/February 2009, pp. 68–70. Reproduced by permission of the author.

Lack of Action

They'd all failed me. They'd failed us all. It had been obvious to everyone for at least 25 years, I thought, that oil would one day run out, that gasoline could not always be cheap, that we could not forever rely on other countries to supply our energy needs. And yet not one damn thing had been done in all that time to guarantee that our energy future would be secure.

I resolved immediately to walk more often to the places I wanted to go. However, living in the Midwest, a place laid out in just such a way as to make driving an absolute necessity for getting around in ordinary life, this was not much of a resolution. In practice, it was impossible to stick with it. I think I walked from home to my friend's apartment twice. I lived too far away from work to walk there even once.

By the way, did I mention this happened 30 years ago?

It was true back in the [19]70s and it is still true today that we don't have an actual energy crisis.

What had so incensed me at the pump that day was that gasoline had shot up to the unheard of price of, I think, 77 cents a gallon. To fill my 20-gallon tank it was going to cost me about what I made in five hours, before taxes. Although it costs me less, measured in hours worked, to fill up today, my minivan tank doesn't hold 20 gallons. I also make a lot more per hour now than I did then with my barely-overminimum-wage job, and I didn't have the expenses of a mortgage, utility bills, and children. What has not changed is that, as I thought 30 years ago and never would have thought would still hold true today (if I hadn't seen it with my own eyes), not one damn thing has been done in all that time to guarantee that our energy future would be secure.

I don't know if it's tragic or just funny that many of the same "fixes" from back then are being suggested for dealing with the current energy crisis. It was true back in the [19]70s and it is still true today that we don't have an actual energy crisis. We're awash in energy and the world is not running out of oil anytime soon. The crisis is in our wallets in that we don't want to (or for many of us, cannot for much longer) spend as much for gasoline (or heating oil, or natural gas, or diesel fuel, or electricity) as we are spending now.

Impact on Personal Finances

This financial crisis is a serious problem for everyone. Though it is easy to pick on people who are driving big SUVs that they suddenly can no longer afford to keep on the road (gas prices went up between 25 and 30% in one year), and laugh at their supposed vanity for owning them in the first place, their money problems propagate through the entire economy. Many can't sell that SUV because the car shoppers who can still afford to own one don't buy used cars, especially when dealerships are trying to unload new ones at bargain prices. Everyone else out car shopping is also looking for something small and energy efficient, which means there is also a shortage of small, energy efficient cars, at least for a while. This means most current SUV drivers are stuck with their big vehicles. This means that money will be spent on gasoline that otherwise would have been spent on clothes, or computers, or refrigerators, or TVs, or on trips to the restaurant....

The nation cannot conserve itself into either prosperity or energy independence.

And this means that even a Green waitress who would never own an SUV, who takes the bus to work, who recycles, who supports saving the rain forests, and who just bought her first little energy-efficient house a year ago—will default on her mortgage when she gets laid off at the restaurant because SUV owners can't afford to eat out as often.

The national economy will continue to weaken one falling liberal and conservative, red and blue state, Green and non-Green, domino at a time.

As I write this in the summer of [20]08, there is talk about reinstating the 55-mph speed limit, a less than perfect idea from 30 years ago. I have myself voluntarily started driving to work with my cruise control set at 60, just to save gasoline. I don't need any more incentive than the high price to give it a try, and I only lose a few minutes both ways by going slower. However, having lived in New Mexico for a few years back in the '80s . . . well, I would never want to try to force someone who has to drive 150 miles to and from work each day to spend an extra half-hour or more on the road. If he's willing to pay for the gasoline so he can continue to spend that half-hour at home, I think that is his right. It is true that if everyone drove 55 on the highways instead of 70 or 75, it would conserve gasoline. But that would only put off the inevitable day of reckoning a short while. And the simple fact is that, if history is any guide, most people will not keep their speed down to 55, and the nation cannot conserve itself into either prosperity or energy independence.

The Downside of Renewable Energy

Just as we heard back in the '70s, once again we are being preached the virtues of clean renewable energy—solar power, wind power, and geothermal power (hydropower, not so much this time around, since no one wants to dam anything anymore). But it is also just as true today that there isn't enough solar, wind, or geothermal power around to meet our energy needs, nor is there likely to be.

No one disputes that a solar power station big enough to supply the energy we get from a typical coal or oil-fired power plant would have to be huge. The back-of-the-envelope calculation is simple. Assume about one kilowatt per square meter as the energy flux for sunlight at the Earth's surface. But at

10% efficiency, ten square meters of solar cells are needed to obtain one kilowatt. To get one gigawatt of power, you need one million times that area. That's a square about two miles on a side. The sun doesn't shine at night, so figure eight hours a day of useful sunshine—now you're up to twelve square miles of solar collector to average one gigawatt of continuous power. Even the desert gets clouds, so we'd better double the size again (and that's being optimistic) to make up for losses we can expect from less than perfect weather. We're now up to 24 square miles of solar collector for that one gigawatt. A typical power plant supplies about five gigawatts. That means we need 120 square miles of solar collector to equal one typical fossil-fuel power plant, a square [that is] eleven miles on a side.

Solar and wind energy will never be more than adjunct sources of energy.

Of course, we're not going to cover 120 square miles with one enormous solar collector. We're going to get our huge solar collector by using a whole bunch of smaller arrays all operating together. There will be space between those units with utility roads, junction boxes, and service buildings. The sun will have to be tracked so the rays will come in perpendicularly. An array with moving parts will tend to break down unless it receives regular maintenance. So either from failure or just to receive scheduled care, some areas of our solar power array will always be out of commission at any given time. To compensate for that, it needs to be bigger still.

I could go on, pointing out that such a huge structure is bound to run into trouble from really bad weather, like tornadoes and hail storms and lightning strikes, and maybe snow and ice storms. Even in mild weather, the damn thing is still going to get dirty and will need to be cleaned frequently. I haven't even touched how much time, manpower, concrete, steel, gasoline and diesel fuel will be needed to build it in the first place.

All of these aspects were known decades ago-they are among the reasons why it was hoped back then that we'd be building solar power satellites in space and beaming the energy back via microwaves.

You can do the same sort of back-of-the-envelope calculation for wind power. No matter how you do it, if you use anything like realistic numbers, you rapidly discover that solar and wind energy will never be more than adjunct sources of energy. I have nothing against either of them in their proper place and under favorable circumstances. Indeed, geothermal power works well for Iceland. But we're not Iceland. Touting solar and wind power as the answer to our long-term energy needs is simply nonsense.

Energy Crisis Solutions

This brings us to the elephant in the room—nuclear power. Just how in the hell did scientists and science fiction writers allow the promise of clean nuclear power to be stolen from us? Because it wasn't so clean? Compared to what? Coal? Oil? The very stuff we're now told is bringing on catastrophic climate change?

Was it the fear of terrorism, that having too many nuclear power plants around would make it too easy to steal nuclear material to make nuclear bombs? But weren't the planes flown into the World Trade Center [in the September 11 terrorist attacks] powered by fossil fuel? Haven't the so-called "rogue states" been making their own nuclear bomb stuff with centrifuges anyway?

Was it the waste that won't be safe for ten thousand years? But haven't other nations been producing waste anyway, and dealing with it? And why does it have to be stored safely for ten thousand years? Can't we settle on a scheme that will serve for fifty or a hundred years, and revisit the problem later, just like we do with absolutely everything else? Are there any "permanent solutions" put forward in 1908 that we still follow today?

My solution to the energy crisis? Nuclear power. We already know how, it is less dirty than fossil fuel, power plants require only acres of land, not square miles, and the US has all the uranium she needs inside her own borders. Forget trying to store the waste forever—settle for a century. We should also drill offshore for more domestic oil since nuclear plants can't be built overnight, and the American automobile fleet won't be all-electric for some years to come. And we can use wind and solar power where it makes sense.

Oh, and since candidates for president can spend a hundred million dollars to get us to vote for them, maybe we can spend the same to undo the 30 years of anti-nuclear nonsense that brought us today's energy crisis in the first place.

6

The Looming Energy Crisis Creates Opportunities for Solar Power

James M. Higgins

James M. Higgins is the George D. and Harriet W. Cornell Professor of Innovation Management at the Roy E. Crummer Graduate School of Business, Rollins College.

Solar-generated electricity is gaining momentum due to innovation in the solar industry, investment in solar energy systems, state requirements to use alternative energy sources, and federal tax incentives that encourage businesses and consumers to use alternative energy sources. Recent innovations in the solar industry field include a process to store solar-generated electricity at a reasonable cost. Within the next ten years, Americans could be using solar-generated electricity to power homes, cars, and businesses.

Solar energy may soon power our homes, office buildings, automobiles, and iPods.

Imagine a solar-powered world. Your home runs on electricity generated by solar panels during the day, and draws from solar energy stored in a fuel cell by night. The windows in your office building collect sunlight and convert it into half of the electricity the building needs; solar electrical power is available at the same or less cost than that provided by coal,

James M. Higgins, "Your Solar-Powered Future," *The Futurist*, vol. 43, no. 3, May/June 2009, pp. 25–29. Copyright © 2009 World Future Society. All rights reserved. Reproduced by permission of The World Future Society.

which is currently used to generate 49% of all the electrical power in the United States. The U.S. power grid is dominated by solar energy.

In this scenario, your electric hybrid automobile delivers the equivalent of 150 miles per gallon but also accelerates from 0 to 60 miles per hour in 4.5 seconds. Your clothing contains solar cells that convert enough sunlight into electricity to run your iPod.

Recent research and development efforts in the solar-energy field will make this world a reality in the next 10 years. Here's how we'll get there from here.

Solar-Energy Conversion Systems

There are three basic types of systems that convert solar energy into electricity:

Flat-panel photovoltaic PV systems. Flat-panel PV systems are the conversion systems that most people associate with solar energy. These use photovoltaic cells, most based on silicon materials, to convert the sun's rays directly into electricity. The industry average for the efficiency of solar-energy conversion into electricity is about 15%, with some cells achieving 20% or slightly more.

Thin-film solar systems. Thin-film solar systems are PV systems based on nanotechnologies that use extremely thin layers of solar-conversion materials applied to a thin layer of flexible backing, usually some type of metal. Thin-film systems are cheaper to manufacture than flat-panel systems, require fewer scarce materials, are much easier to install, and require less physical vertical space per installation since they are flat.

Thin-film solar materials can deliver virtually the same efficiencies as most flat-panel systems but at about 20% of the cost. Representative of thin-film manufacturing companies is Nanosolar, which has achieved rates of 14% conversion efficiency in its marketed products. Nanosolar has lowered the costs of its products more than the average thin-film solar firm because it has developed a printing-press-like process that enables mass production of its products at very low per-unit costs.

Even higher efficiency rates are coming to the thin-film marketplace because 20% efficiencies have been achieved by researchers at the U.S. National Renewable Energy Laboratory. Thin-film solar is itself a recent major innovation in the energy industry and one of the primary reasons that solar will soon eclipse coal as a major energy source.

Concentrator solar systems. There are two types of concentrator solar-energy conversion systems. The first of these, thermal-concentration systems, focus the sun's rays on heat-retaining media such as water to generate very high levels of heat to create steam to drive generators that produce electricity. Think of a thousand large mirrors surrounding and focused on a central elevated tank of water or other heat-retaining medium connected to a generator or turbine, and you have the idea.

The second type of concentrator, concentrator PVs (CPVs), focuses the sun's rays on PV cells to generate electricity directly. This type of concentrator saves money because fewer expensive and increasingly scarce PV materials are needed. In addition, less space is generally needed for concentrators than for flat-panel installations. The number of CPVs being used is increasing rapidly, and many variations of CPVs . . . coming into the marketplace . . . offer different cost and efficiency performance levels.

Recent developments in all three of these conversion systems will contribute to the future dominance of the energy industry by solar. Let's now examine the solar future.

A New Dynamic Duo: A Solar And Fuel-Cell Combination

There have been several very significant innovations in the solar-energy field in recent months. Probably the most significant of these is the development by MIT [Massachusetts Institute of Technology] scientists Daniel Nocera and Matthew Kanan of a process for storing solar-cell-generated electricity in a fuel cell for later use . . . when the sun is not shining.

The key to the success of their process was their formulation of a catalyst, which makes this fuel-cell energy-storage process possible. As Anne Trafton of *MIT News* notes: "Homeowners could use their solar panels during the day to power their homes, while also using the energy to split water into hydrogen and oxygen for storage. At night, the stored hydrogen and oxygen could be recombined using a fuel cell to generate power while the solar panels are inactive."

Nocera believes that products based on this technology could be on the market within 10 years. If their invention receives significant funding, it could be much sooner. The implications of this discovery are huge:

- The home or business storage of solar-generated electrical energy at a reasonable cost would be possible for the first time.

- Electric and electric-hybrid automobiles would have a power source that would not drain the national power grid. Imagine what 200 million electric and electric-hybrid automobiles would do to power-grid demand if they were all plugged in at about 10 P.M. every night. This invention helps solve that problem by taking millions of these autos off the power grid. This helps pave the way for the mass utilization of plug-in electric automobiles.

- The national power grid would need fewer new large power plants to meet growing customer demand, since its utility members could buy the surplus energy from homeowners with solar energy systems. As it stands now, in a business-as-usual scenario, many more large power plants will be needed to meet the electricity needs of an expected 42% increase in U.S. population by 2050.

- When combined with other inventions discussed in this article, solar-generated electricity becomes relatively inexpensive and comparable in cost to coal.

Opening the World's Windows to Solar Energy

According to *Science Daily*, there are some 76 million residential buildings in the United States and nearly 5 million commercial buildings. Buildings in the United States produce about 35% of all of U.S. carbon dioxide emissions.

"Buildings account for about 40% of the world's total energy consumption and 65% of electricity usage in the U.S.," according to Nick Beglinger and Tariq Hussain, writing in *Strategy + Business*. Reducing the energy used by buildings has therefore become a critical issue in energy conservation.

But what if we could also use those same buildings to generate much of the electricity they require? Well, we will soon be able to do so. In the future, the majority of the windows on most of those 81 million buildings can be used to collect solar energy and convert it to electricity.

An MIT team led by engineer Marc Baldo has developed "solar concentrators," which move usable sunlight to the edges of windows, where conversion materials will create electricity from it. Thus, most of the windows' surfaces still serve the normal function of windows.

Baldo points out that this system can also be used to increase the energy efficiency of normal PV solar collectors by as much as 50%. He believes that this collector system will be on the market within three years, and will be especially effective in large office buildings with numerous large windows.

Your future will be affected greatly by these solar concentrators:

- The windows where you work, live, shop, or enjoy recreation will be busy working to lower your cost of living, and you still can use them as windows.

- Buildings, including homes, will be designed around window-based solar-energy-conversion systems as well as solar-conversion systems on roofs or side walls.

- Many, if not most, buildings will be retrofitted to accommodate these solar concentrators. The implications for the construction industry are huge.

Solar-Generated Electricity at the Same or Less Cost Than Coal

Half of the electricity in the United States is generated by coal, the most greenhouse-gas-emitting major energy source. Coal has historically been abundant and cheap, but its price is rising with the increased demand in developing economies, primarily China and India. Solar-generated electricity is now beginning to become cost competitive with coal-generated electricity. And once installed, solar-energy conversion systems do not emit greenhouse gases.

Four key factors will make solar-generated electricity competitive with the cost of coal-produced electricity in the next year or two: the innovations that are taking place in the solar industry, the significant amount of money being invested to bring scalability to solar-energy-system manufacturing, state requirements that alternative energy sources constitute a specific percentage of a utility's total mix of energy sources, and tax incentives from the federal government to encourage consumer and business investment in solar and other alternative energy sources. With the new [President Barack] Obama administration, there is a potential fifth factor: a federally mandated percentage for alternative-fuel power generation by 2020.

Recent Innovations in Solar Energy

With the emergence of low-cost thin-film solar, the rapid evolution of concentrator photovoltaics, and the increased capabilities of thermal solar concentration systems, the solar energy industry has begun to shine brightly. Additional recent innovations that will have significant impacts on the energy industry include:

- IBM's concentrator solar cell. What has historically stymied progress in using PV solar concentrators is the high levels of heat that they generate. Normal PV cells melt under such heat. IBM has developed a way to cool solar cells so that they can withstand extremely high levels of heat by using the same technology used to cool computer chips. IBM's cooling process makes this particular concept doable. Concentrators based on this type of technology are going to be cost competitive with coal as an electricity-generating energy source and much more competitive than wind-power systems.

Solar-generated electricity is now beginning to become cost competitive with coal-generated electricity.

- Another concentrator system, developed by researchers Anna Dyson and Michael Jensen of Rensselaer Polytechnic Institute and being tested at two commercial sites, uses a concentrator lens with concentric grooves to focus the sun's light and heat on postage-stamp-sized high-tech solar chips. The use of both heat and light may allow efficiencies as high as 80%.

- In plastic-based solar-energy conversion systems, organic materials are manipulated at the nano [submicroscopic] level to provide paths for separating positive and negative charges and then turning these into electricity. Plastic conversion systems are very inexpensive, but they are also much less efficient than other conversion systems. However, where space and efficiency are not issues, plastic converters are just what are needed. And they are so practical in terms of application that someday you might even be able to create your own solar panel using an ink-jet printer to print on thin sheets of plastic, paint a large wall with plastic-based solar paint, or paint more advanced plastic solar materials on the roof of your car and create enough energy to propel it.

- Solar cells made of "nano flakes," developed by researchers at the University of Copenhagen, use a pure crystalline structure that doubles conversion-efficiency levels.

- Solar start-up CaliSolar's "crystallization" process enables solar conversion systems to use less pure silicon materials because they have been redesigned to herd all of the impurities together, thus avoiding the problems normally associated with such impurities. This significantly reduces costs without lowering efficiencies.

- Prism Solar developed solar cells that use holograms to divert less-useful sunlight away from solar cells, thereby increasing conversion efficiencies.

The solar industry is going to expand rapidly, creating new jobs in hundreds, maybe thousands, of new businesses.

These and other R&D [research and development] efforts in companies and universities will help drive down the costs of solar-generated electricity to highly competitive levels in the near future. The solar industry is going to expand rapidly, creating new jobs in hundreds, maybe thousands, of new businesses—solar construction, service, education, and more.

Investments, State Requirements, and Tax Incentives

Innovation, while the cornerstone of change, is not enough by itself to propel solar energy to competitiveness. Four additional factors are critical.

1. Investment in scalability. The energy industry has long been dominated by large firms that benefit from economies of scale and whose primary interests have been maintaining the status quo. The "greening" of energy production has so far been somewhat of a cottage industry, led by a host of small and medium-sized entrepreneurial companies backed by venture capitalists. Now, a growing number of large, non-energy-based companies like Intel and IBM have been making huge investments in solar energy. This investment will increasingly drive costs down through economies of scale in manufacturing.

2. State requirements. California and other states have mandated that electric-utility companies use alternative energy sources to generate electricity, and solar is increasingly the alternative energy of choice. In California, Pacific Gas & Electric states that it already achieves 24% of its sources from eligible renewable resources. Such huge commitments by large electrical utilities will enable smaller and medium-sized solar firms to reach scalability, as well as encourage further development of industry capabilities.

3. Federal tax incentives. Congress uses tax incentives to encourage consumers and businesses to install solar and other alternative energy systems. These inducements have played a major role in helping build the solar and other alternative-energy industries. Continuing these

policies over the next decade will be critical to developing solar industries further.

4. A federally legislated percentage for alternative fuel power. President Barack Obama has indicated his support for a substantial reduction in U.S. greenhouse-gas emissions by 2020. Along with plug-in hybrid automobiles, one major way to achieve this would be a requirement for 25% alternative-fuel power generation by 2020.

Solar is increasingly the alternative energy of choice.

For all of these reasons, I believe solar will begin to dominate the national power grid within the next 10 years. Currently, large sections of arid and desert lands in eastern California and other states are being sold or leased for solar farms. The most likely scenario will find solar energy dominating the power grids in western and southern states first, taking advantage of the sun's intensity in these regions, then spreading to power grids in the rest of the country as solar-conversion systems continue to improve in conversion efficiency and versatility.

More Solar Power Under Your Control

Ten years from now, power generation will be much more widely distributed. Homes and businesses alike will install solar-energy conversion systems for most—if not all—of their electrical needs. Centralized power from electrical generating plants will still be needed, but in the future they will largely be a backup for distributed electrical generating systems rather than the sole provider of electricity. Moreover, these traditional power plants will predominantly rely on solar energy rather than coal, gas, and oil.

Even our clothing will offer nongrid power opportunities for solar energy, thanks to breakthroughs in nanotechnology. By integrating solar—collection materials into fabrics, we'll have shirts and sweaters that can power our iPods or cell phones, and window drapes that help power our homes and offices.

The rise of new solar technologies will [harness] the power of sunshine in a host of new ways, ultimately giving you the means to become the manager of your own solar-powered lifestyle. You'll hold the sun and its power in your hands.

7

Global Food Prices Have Increased Due to the Energy Crisis

Jacqui Fatka

Jacqui Fatka is an Iowa State University graduate and is a Feedstuffs staff editor.

As the price of oil increases, the price of food increases. Currently, corn-based ethanol is the only economical alternative to gasoline. U.S. farmers are producing more corn than ever to accommodate the growing demand. Corn-based ethanol is not responsible for increasing food prices; oil shortages are to blame.

First, overproduction of corn was blamed for the increasing use of high-fructose corn syrup and obesity. In less than a decade, the tables have turned, and now, corn is accused of causing world hunger.

While many questions remain about biofuels, the energy crisis is at the heart of rising food prices and what also first led U.S policymakers to look to alternative energy sources.

The ethanol industry is the first to admit that corn-based ethanol is not the perfect biofuel, nor [do producers] plan on converting the entire corn crop to ethanol. Corn-based ethanol, while not a silver bullet, is the foundation upon which the next generation of "advanced biofuels" is being built.

Jacqui Fatka, "Commitment to Ethanol Stands: Corn-Based Ethanol Helps Lay Groundwork for Future Ethanol Feedstocks to Help Alleviate Soaring Oil Prices," *Feedstuffs*, vol. 80, May 19, 2008, p. 9. Reproduced by permission.

Brazil's ethanol is derived from sugarcane, one of the least expensive feedstocks. However, in the U.S., studies have shown that because of domestic sugar programs, sugar is not the least expensive feedstock.

The energy crisis is at the heart of rising food prices.

Today, corn is the cheapest feedstock to convert into ethanol. Although research exists to convert cellulosic materials such as corn stover (corn's stalks and leaves left over after harvest), corn cobs, switchgrass and wastes into ethanol, the technology has yet to bring costs down to economical levels.

The December 2007 energy bill included a 15 billion-gallon mandate for corn-based ethanol by 2015 and also expanded the mandate to 60 billion gallons by 2030, with the extra coming from non-food-based sources.

Rep. Lee Terry (Republican, Nebraska) said his vision, as with many legislators, is that biofuels are going to have to be a part of the domestic energy portfolio, but that portfolio has to be "varied" and "diverse."

Research needs to be dedicated to transitioning from first-generation biofuels to the second, third and fourth generations, where the process will be efficient and affordable, he said recently at a House hearing.

Earlier this year, the Department of Energy granted funding to six cellulosic ethanol pilot plants, investing up to $385 million over the next four years. When fully operational, the biorefineries are expected to produce more than 130 million gallons of cellulosic ethanol per year. This compares to the nearly 11 billion gallons of corn-based ethanol production capacity planned to come on line by the end of the year.

The disadvantage of using cellulosic material is that a pound of cellulose yields only 70% as much ethanol as a pound of corn kernels, creating the need to transport, store and handle huge volumes of product.

Food Prices

A $3 box of corn flakes contains 15 ounces of corn that cost 8 cents when bought from the farmer.

Meanwhile, U.S. farmers last year grew a record-setting 13.1 billion bushels of corn on 85 million acres. Of that, 22% went to make about 7 billion gallons of ethanol. That still left enough corn to supply the domestic market, increase exports to record levels and stockpile a 10% surplus.

While corn prices have had a small role in increasing food prices, Rick Tolman, chief executive officer of the National Corn Growers [Association], said the industry is "closer to Little Bo Peep than the ax murderer."

Domestic food inflation [in the United States] is up 4.5% this year, while global food inflation is up approximately 43%. White House chairman of the Council of Economic Advisers Ed Lazear puts ethanol's contribution to domestic food inflation at only 0.25% and 2–3% of world food inflation increases.

Lazear explained that increasing demand in emerging markets such as China and India account for about 18% of the rise in global food prices.

A $1/gallon increase in the price of gas has three times the impact on food prices as a $1/bushel increase in the price of corn.

A new report from *Biofuels Digest*, "Meat vs. Fuel: Grain Use in the U.S. & China, 1995–2008," concluded that increased Chinese meat consumption requires six times the amount of grain compared to what the U.S. ethanol industry uses annually.

Another key component is rising energy costs, which adds costs to processing, packaging and transporting food items. In addition, adverse weather-related events this year in Australia, China and many Eastern European countries that produced low crop yields took prices sharply higher.

A $1/gallon increase in the price of gas has three times the impact on food prices as a $1/bushel increase in the price of corn.

Lower Gas Prices

According to a study by Iowa State University, growth in ethanol production has actually caused retail gas prices to be 29–40 cents/gallon lower than they would otherwise be.

On the coasts, where gas prices are especially high, drivers are saving an average of $142 a year on regular unleaded. In the Midwest, they are saving $241 a year.

The researchers found that on the coasts, drivers save an average of 23 cents/gallon.

The savings are significant in other parts of the country, too: On the Gulf Coast, it is 25 cents/gallon; in the Rocky Mountain states, it's 17 cents, and in the Midwest, where U.S. ethanol production is concentrated, drivers save 40 cents/gallon.

Also, Merrill Lynch [a financial management and advisory firm] analysts noted that oil and gas prices would be 15% higher if not for the availability of ethanol. At [2008] national average pump prices, this puts ethanol's value at around 50 cents/gallon to American consumers.

Ethanol is the only alternative to gasoline available today, and its production is generating economic benefits for the [United States] while reducing the need for imports of oil and gasoline, often from unstable or openly hostile sources.

Here's the Point

Corn-based ethanol has come a long way since its inception [during the 1970s]. Although it is not the "silver bullet" in addressing rising energy costs and reducing dependence on foreign oil, it's a solution that is here now, is available in the current infrastructure and is making a difference in the price of fuel.

The ethanol boom has brought renewed economic strength to rural areas [by] adding jobs and stimulating economic growth. Studies also show that the net cost to consumers and taxpayers is zero to slightly positive.

Corn producers stepped up to the plate in 2007 to produce a record corn crop of 13.1 billion bushels. Technological advancements are helping producers grow more on fewer acres.

The role corn prices play in food prices has been minimal. However, rising oil prices remain at the center of food price increases, with oil rising from $35 a barrel in 2005 to more than $120 this year.

8

Fear Blocks Opportunities to Resolve the Energy Crisis

James Lovelock

James Lovelock is an independent scientist, author, researcher, environmentalist, and futurist who lives in Cornwall, in the southwest of England. He is known for proposing the Gaia hypothesis, in which he postulates that the Earth functions as a kind of superorganism.

The world is increasingly dependent on oil and gas. Although nuclear energy is a viable solution to the energy crisis, energy companies and politicians are using fear to block the use of nuclear energy. The use of nuclear power would cut the profits of energy companies and hinder the ability of politicians to accomplish certain agendas.

Normally the media can smell a rat better than a hungry terrier [can], and I was slightly surprised that they did not wonder more about the murder of the Russian dissident Alexander Litvinenko in 2006 in London.

He was cruelly poisoneds by a few hundred nanograms of the radioactive isotope polonium-210. When swallowed it soon finds its way to every cell of the body, where it emits helium atoms that plough through the vital structures. An evil way to kill someone: a slow, unstoppable, tortured death.

There is ample evidence that the agents of the murder were Russian, and the container of the radioactive element was leaky enough to leave a trail from the airliner that brought the assassin to London to the hotel where the poison was added to a cup of the victim's tea.

What an opportunity was missed by some imaginative journalist or thriller writer, to have set a scene somewhere in Moscow with a cast of professionals from security agencies or energy corporations.

It would be naive to expect energy companies to stand aside and see their profits cut.

Someone says: "You realise, do you, that a poisonous dose of polonium-210 will cost about $10 m [million]? Why not use ricin—we know that that's a reliable poison and a lot less visible to the media? Moreover, it will cost less than $1." Another bureaucrat adds: "Yes, and to make the polonium we have to seek time on a reactor which is already fully occupied with other important tasks." At which a senior manager intervenes.

"Gentlemen," he says, "the purpose of this action is not merely to punish a traitor—and that alone needs visibility and media amplification—but more importantly to keep the West frightened of all things nuclear. Our future as a world power depends on our ability to make them wholly dependent on us for their supply of oil and gas; their use of nuclear energy would free them of this dependency and we could lose our ability to make the world go the way we wish. Ten million dollars is nothing in that cause." This scene is no more than a figment of my imagination, but it grows more credible as we move further into the 21st century, when political power and business opportunity will more and more be linked to energy supply. It would be naive to expect energy companies to stand aside and see their profits cut by inexpensive nuclear energy, and the same must be true for the thwarting of national aspirations.

Uncertain Future

Our greatest future need in the UK will be a secure supply of food and energy. Soon the growing appetite of the world for both, and the worsening climate, will make the supply from abroad increasingly more expensive, and we will be driven more and more to produce food and generate electricity from our own resources.

We in Britain are no longer a major manufacturing nation, and [we] may have to leave the engineering development of our energy supply to those nations better equipped to do so. The worst of all possibilities would be for us to become the test-bed for unproven technology, and this is what is happening now with wind turbines.

We should regard nuclear energy as something that could be available from new power stations in five years and could see us through the troubled times ahead, when the climate changes, and there are shortages of food and fuel, and major demographic changes [occur].

The only thing that stops an immediate build of new nuclear electricity is legislation put in place by previous governments and unreasoning fear.

[People] in Britain should think of the troubled years of the 1970s and early 1980s, when industrial conflict over coal threatened electricity supplies.

It was the availability of nearly 30% of the electricity we used from nuclear energy that sustained the nation and stopped the quarrel turning into a civil war. The only thing that stops an immediate build of new nuclear electricity is legislation put in place by previous governments and unreasoning fear.

The Truth About Nuclear Power

There are now more than 440 nuclear power stations in the world, producing 17% of all the electricity used, about the same percentage as hydroelectricity.

Other sources of renewable energy—biofuels, wind, etc—produce only 2%. The safety record, their cost and the local acceptability of these fission-powered stations make them the most desirable of all sources. So why in the First World do we still persist in the falsehood that they are uniquely dangerous? I think we fail to welcome nuclear energy as the one good and reliable power source because we have been grievously misled by a concatenation of lies. Falsehood has built on falsehood and is mindlessly repeated by the media until belief in the essential evil of all things nuclear is part of an instinctive response.

It is often said that nuclear waste is uniquely deadly and will persist for millions of years and poison the global environment. All pollution by chemical elements persists. Lead pollution from a mine, smelter or factory where it is made into things lasts forever; the same is true of mercury, arsenic, cadmium and thallium: these toxic elements are permanently with us. What is remarkable about nuclear waste is that it fades away. In 600 years the high-level waste from a nuclear power station is no more radioactive or dangerous than the uranium ore from which it originated. More importantly, there is hardly any nuclear waste to worry about. The yearly output of waste from a 1,000 MW [megawatt] power station would fit in a London taxi.

Belief in the essential evil of all things nuclear is part of an instinctive response.

Even government committees such as the Committee on Radioactive Waste Management propagate nuclear falsehood: One of its representatives said there is enough nuclear waste in Britain to fill [London's Royal] Albert Hall five times over. In fact, after 40 years of generating nuclear energy, there is barely enough to fill one Albert Hall.

Compare this with the mile-high mountain, 12 miles in base circumference, of solidified carbon dioxide that the world makes every year. The nuclear waste is a minor burial problem, but the carbon dioxide waste will kill us all if we go on emitting it.

The Truth About Wind Energy

In addition to the negative propaganda directed at nuclear energy, there are almost as many untruths propagated about the favourable qualities of wind energy. Were these wind farms truly efficient and capable of resolving our power needs, I might be persuaded to grit my teeth and endure their ugly intrusion, but in fact they are almost useless as a source of energy.

It would take a vast area of countryside to provide enough land for a one-gigawatt wind-energy source. The wind blows only 25% of the time at the right speed to generate a useful quantity of electricity; therefore this monster would need the back-up of a near-full-size fossil fuel power station to supply electricity whenever the wind blew too much or too little.

Take, for example, the British intention to build the world's largest wind farm in the Thames estuary, which would have 341 turbines occupying an area of 90 square miles. It is claimed to be a one-gigawatt project and therefore equal in output to a typical nuclear power plant. In the hype attending it is the claim that it will provide enough electricity for one-third of London's homes and save the emission of 1.9 m tons of carbon dioxide. It sounds good until you realise that a full-size, presumably coal-burning, power station, emitting copious amounts of carbon dioxide, will have to be built to back it up when the wind does not blow.

Its real averaged output would be only 400MW of electricity. If it were steady, which it would not be, it would be enough for 830,000 homes each consuming 4,200 kWh [killowatt hours] yearly. I am glad the oil company Shell had the wisdom, despite subsidies, to pull out of this flawed project.

To survive on these islands with a future population perhaps as large as 100m requires a constant and reliable source of electricity from indigenous fuel. It would be madness to attempt it without nuclear energy.

It is sad that so many of the green movement and their intellectual followers still oppose nuclear on grounds as insubstantial as a fear of hellfire and Satan.

9

The Peaking of Oil Is Causing Tension Between Nations

Tony Black

Tony Black writes for Canadian Dimension, *a magazine of liberal political thought.*

The peaking of oil is a result of declining rates of discovery and extraction, along with increased oil consumption. As oil becomes scarce, tensions are rising between oil-rich countries and countries without vast oil resources. [Former U.S. president] George W. Bush's "Axis of Evil" consists of nations with major oil reserves and nations controlling sea-lanes vital to the transportation of oil.

To the long list of wearying apocalyptic scenarios facing the future of humankind must, unfortunately, be appended yet another. One which, nonetheless, at least has the merit of focusing attention by virtue of its sheer immediacy.

According, then, to many of the world's most prestigious (independent) oil geologists and institutions, not only is the era of cheap oil now almost certainly at an end, but by the end of this decade—and likely before—the price of a barrel of oil will rise well past $100, and will continue to climb quickly and inexorably thereafter. One need not be a rocket scientist to begin to grasp the staggering implications of this for industrial civilization.

But before we do that, best to start at the beginning. . . .

Tony Black, "The Party's (Almost) Over: 'Peak Oil' and the Coming Global Energy Crisis." *Canadian Dimension*, vol. 39, September–October 2005, pp. 14–16. Reproduced by permission of the author.

The Problem

In 1956, the geologist M. King Hubbert predicted that the U.S.'s oil production would peak in 1970, and thereafter begin an accelerating decline. Despite being derided at the time by many industry analysts, his predictions proved entirely accurate. This, for the very good reason that he had correctly figured out that oil extraction follows a statistical model known as a bell curve.

The era of cheap oil [is] now almost certainly at an end.

Oil production starts off slow, picks up more and more, reaches a peak—when half the oil has been pumped out—and then begins to fall as it had risen, that is, faster and faster. In the latter stages, all sorts of strategies are used to maintain pressure at the well-head, including pumping the reservoir with water and gas. Still, it's a case finally of diminishing returns. Less and less oil at greater and greater cost.

But, you might ask, what difference does it make if a particular oil field runs dry? Aren't new ones being discovered?

Aye, there's the rub. In fact, many of the world's largest oil fields are now decades old and are depleting rapidly. Thus, in the mid-[1980s] the North Sea used to produce 500,000 barrels per day. Today it produces 50,000. Prudhoe Bay in Alaska once gushed out over 1.5 million barrels per day, but in 1989 it peaked and now gives only 350,000. The huge Russian Samotlor field used to account for 3.5 million barrels per day. Now its ledger tallies a mere 350,000.

Even the largest oil field in the world, the giant Ghawar of Saudia Arabia, is showing distinct signs of having peaked. The Saudis must now inject some seven million barrels of water per day into the reservoir just to maintain well-head pressure.

As for new discoveries, there have been no new large discoveries in over two decades. The 14 largest oil fields average over forty years old. Indeed, Dr. Colin Campbell, former chief geologist for Shell Oil, has stated that the discovery of major new oil reserves "peaked in the 1960's" and that, "We now find one barrel for every four we consume."

According to the best estimates, by such as the Petro-consultants of Geneva, the French Petroleum Institute and the Colorado School of Mines, global "peak oil" will likely be upon us sometime before 2010. Such estimates, of course, fly in the face of [ones] by the U.S. Geological Survey and the International Energy Agency, both of which acknowledge "peak oil," but which project its onset to somewhere between 2015 and 2040. Most of the independent experts, however, argue that these latter figures are mistaken on at least three counts.

The first is that oil reserve figures have been inflated for years for purely speculative and ideological reasons, essentially to maintain investor confidence. A stark illustration is provided by the scandal of Shell Oil, which was recently forced to revise its reserve estimates downwards by over twenty per cent.

The second involves reliance on so-called "non-conventional" oil supplies, including tar sands and oil shales. Unfortunately, these require massive amounts of energy to extract. And not just energy. For every barrel of oil from tar sands, 400 to 1,000 cubic feet of gas are required. For every barrel of oil from shale, one to four barrels of water are needed. Moreover, the tailings and detritus left behind are literally mountainous, an environmental nightmare just waiting to explode.

They also take time to bring on line. It is expected, for instance, that Canada will, by 2030, be producing no more than four million barrels of oil per day from its Athabascan oil sands. This would amount to little more than three per cent of the (now widely agreed upon) estimated global need of 120 million barrels per day (mb/d) by 2025 to 2030. (Present usage is about 84 mb/d).

Finally, the term "peak oil" refers not just to the problem of declining rates of discovery and extraction, but also to the compounding problem of increasing rates of consumption. China, for instance, has recently surpassed Japan as the world's second-largest importer of oil, imports which are rising by nine per cent a year. India's thirst is also exploding.

In short, the gap between demand and supply is already beginning to yawn. Once the downhill side of the bell curve sets in, this gap is likely to spread into a chasm. And the consequences of that are little short of catastrophic.

$150 per barrel of oil will profoundly affect everything from our suburban, commuting way of life right through to our industrial and international modes of agricultural production and distribution.

Goodbye Growth, Goodbye Globalization

For the past century, the world's industrial societies have enjoyed between two and seven per cent annual economic growth. This growth has been almost entirely fuelled by a bonanza of cheap, easily extractable, high-grade oil. Virtually every aspect of our modern industrial economy relies on it, including global transport of goods, commercial air travel, gasoline for cars, the lubrication of industrial machinery, the generation of electricity and the production of plastics, fertilizers and pesticides. The consequences of, say, $150 per barrel of oil will profoundly affect everything from our suburban, commuting way of life right through to our industrial and international modes of agricultural production and distribution.

And that's just to begin with. World economic growth will likely become a quaint curiosity consigned to the history books. Permanent recession, if not outright depression, could well become the norm. Unemployment will undoubtedly skyrocket, as will world political instability. Driving a car? Taking a plane? These will gradually, incrementally, become elitist activities reserved for the few. Agriculture will need to become regionalized and localized, just as our whole modern way of life will need be re-engineered to accommodate our changed energy circumstances.

Given all this, one might be hard pressed to comprehend how officialdom and the major media have failed both to recognize and/or respond adequately to the problem.

The narrow answer to this conundrum is that official reports have, for years, simply concentrated on global oil reserves (which are still quite extensive), rather than on the projected future gap between production and consumption. Thus, the world will continue to "enjoy" significant reserves far into the future. Oil will not, all of a sudden, simply run out. But the gap between supply and demand will grow quickly. Price volatility will follow suit.

The broad answer is, of course, that certain government bodies have already responded to the problem.

Resource Wars of the 21st Century

That the U.S. invasion of Iraq was mostly about oil will come as no earth-shattering revelation to many. What is perhaps less well appreciated is how completely the list of nations [constituting George W.] Bush & Co.'s "axis of evil" extend along an arc that maps virtually one-to-one both to the world's major reserves of oil, and to the strategic chokepoints and sea lanes vital to its transport and distribution.

U.S. threats against Iran, for example, not only target one of the world's largest reserves of oil and the critical Strait of Hormuz, linking the Persian Gulf to the Arabian Sea, but also China itself, which has signed major oil deals with Iran. China's energy supplies and supply lines are further threatened by the U.S.'s sudden humanitarian interest in the Darfur region of the Sudan, where China also has major oil concessions; and by U.S. actions against Venezuela, with whom China has recently negotiated a major bilateral energy deal; and by Washington's recent naval and troop deployments to the Strait of Malacca, which controls access to the South China Sea. . . .

That the U.S. invasion of Iraq was mostly about oil will come as no earth-shattering revelation to many.

Analogous considerations apply to a host of other countries from Yemen and Somalia (straddling the vital oil-transit strait to the Red Sea) to Algeria (90 per cent of whose oil goes Europe) to a string of countries girding the belly of Russia.

With respect to the latter, a very interesting contest is shaping up in the Caspian Sea basin. The American-backed Baku-Ceyhan pipeline (bringing oil westward and bypassing Russian pipelines) has recently been completed and the Kazakhstan-China oil pipeline is expected to come on line near the end of the decade. Meanwhile, though China and India have squared off in Angola, Indonesia and the Sudan, they have also engaged in intensive negotiations with respect to their energy security needs in Central Asia. There is talk of extending the Iran-Pakistan-India gas pipeline to eastern China.

The recent spate of "velvet" revolutions in Central Asia are clearly tied in with the American attempt to control this vital, energy-rich region. Overall, Russia, China and India are, in a very real sense, strategic allies who share common cause both in frustrating Washington's attempt to isolate Iran, and in securing sources and transit routes from the oil-rich Middle East and Central Asia to oil-deficient East Asia.

In summary, the struggle, the war if you like, for the globe's limited energy resources has, in other words, already begun.

Sleepwalking into the Future

Are alternative fuels and alternative technologies the answers to the coming crisis?

The short reply is, possibly yes, probably no. Gas, for instance, is being touted as a clean alternative fuel. Unfortunately, gas is not only running out fast, but its depletion curve has the peculiar property that, when it does run out, it stops suddenly, with little warning. Moreover, to tap, say, Russia's great reserves of methane, it has first to be liquefied, at great cost, for transportation.

In summary, the struggle, the war if you like, for the globe's limited energy resources has, in other words, already begun.

But then what about the renewables? Well, bio-fuel requires fertilizers—which require oil. In any case, land is needed, it's hardly necessary to add, to grow food. Hydrogen is clean, but it takes a lot of energy to produce it. A hydrogen economy is, really, no more than a fantasy. Wind and solar power are unlikely to ever significantly substitute for fossil fuels, at least given our present energy-intensive lifestyles. Nuclear is also being touted once again, but its capital and decommissioning costs are exorbitant, and its . . . waste [is] problematic. Moreover, uranium is also a limited resource.

This leaves only a multi-faceted approach in which our remaining reserves of oil and gas are used to transition to a combination wind, solar and coal energy economy, while we simultaneously embark on mass-scale efficiencies, conservation and downscaling of our energy usage. It will be a different world. This in the best of scenarios, where immediate action is taken to transition to the post-oil era, a period many experts have said would take a minimum twenty-year effort.

But given how little open recognition there is of the problem, let alone any concerted worldwide policy to implement change, it seems rather more likely that, intoxicated for an entire century with the heady brew of cheap energy, the party will continue to rage on. That is, until the fridge runs dry, the tempers wear thin and the furniture is tipped, unceremoniously, on end.

10

Industrialized Nations' Economies Could Suffer If the Oil Supply Is Blocked

Gulf Daily News

The Gulf Daily News *is an English-language newspaper published in the Kingdom of Bahrain by Al Hilal Group.*

The Strait of Hormuz is used to transport a significant portion of the world's daily consumption of crude oil. In August 2008, the world took notice when Iran threatened to close the Strait of Hormuz. It is speculated that if such an event occurred it would have an adverse effect on economies throughout the world.

They are the arteries of the world, the waterways that keep our global economy alive.

Two-thirds of all seaborne trade [and half of the world's daily diet of oil] passes through these six narrow choke points.

Close [down] just one, for just one day, and the heartbeat of the industrial world would falter.

Block it for a month, and entire economies could collapse.

It takes a dramatic event to remind us of the importance of these vital thoroughfares.

When Egyptian leader Gamal Abdel Nasser nationalised the Suez Canal in 1956, Britain and France panicked.

They had good reason. Almost all the traffic between Europe and Asia then passed through the waterway, which cut 4,400 nautical miles off the old sea route from London to Bombay.

The very idea of the canal falling into local hands appalled the former colonial powers.

Yet even this channel, one of the engineering wonders of the modern world, pales in importance beside the Strait of Hormuz.

Two-thirds of all seaborne trade [and half of the world's daily diet of oil] passes through these six narrow choke points.

The Strait of Hormuz and Oil

Like the man-made Panama Canal, [which] cuts through the Americas, the way through Suez is a passage of convenience—there are alternative ocean routes. By comparison, the natural waterway of Hormuz is the only viable path today for most of the oil that flows out of the [Persian] Gulf.

As much as a fifth of the world's daily consumption of crude oil would be choked off if anything blocked the Strait of Hormuz.

And that is precisely what Iran is threatening to do.

Iranian Revolutionary Guard commander Mohammed Ali Jafari is known for his bombastic statements.

So when, on August 4 [2008], he claimed Iran had tested a "unique and completely Iranian-built" weapon that could close the Strait of Hormuz "easily and on an unlimited basis", military experts did not start ripping up their manuals.

The best guess of [*Jane's Defence Weekly* magazine's] defence analyst Doug Richardson is that Iran has just taken delivery of a new consignment of Russian anti-ship missiles.

Still, Tehran knows how to provoke a reaction. Ever since the tanker wars of the 1980s, the Iranian regime has periodically threatened to close the strait in response to outside pressure.

In reality, nothing of the sort has ever happened.

But this does not stop the Western press and think tanks from indulging in a little Suez-style hysteria.

And, as in the 1950s, they do have good reason. As Henry Pokolski, executive director of the right-leaning Nonproliferation Policy Education Centre points out, closing the strait "could reduce US gross domestic product by seven per cent after 30 days, which would be an extraordinary blow".

The natural waterway of Hormuz is the only viable path today for most of the oil that flows out of the Gulf.

The Strait of Hormuz and the Global Economy

There is no doubting the importance of this narrow waterway in world affairs.

A horseshoe-shaped body of water that stretches between Iran and the northern tip of Oman, the Strait of Hormuz is the only way in and out of the Gulf. On a typical day, around 50 tankers carrying between 14 million and 17m [million] barrels of oil and oil products pass through the 180 km-long strait—roughly 40 per cent of the world's internationally traded supplies.

Dozens of freighters ply the same route, carrying food and consumer goods into the ports and free zones of Saudi Arabia, Bahrain, the UAE [United Arab Emirates], Qatar, Kuwait—and Iran.

Jebel Ali alone is now the largest port between Rotterdam and Singapore, handling some 9.9m 20-foot container units last year—equivalent to half the maritime traffic between Europe and Asia.

"About half of what goes into the Gulf is for transshipment and another half is local imports—although a lot of that goes into free zones and is exported again in another guise," says Neil Davidson, of Drewry Shipping Consultants.

In other words, not only do Gulf economies depend on this regular cargo traffic through the Strait of Hormuz, but the outside world does too.

Closing the strait "could reduce US gross domestic product by seven per cent after 30 days, which would be an extraordinary blow."

Then, too, there is the indirect traffic between the Gulf and the global economy. According to *Jane's Intelligence Review* of London: "If this choke point was closed for an extended period, the economies of the Middle East would suffer significantly and this would generate severe economic dislocation around the world.

"Millions of guest workers in Gulf states from developing countries could also be left unemployed, leading to greater poverty in South Asia and East Asia."

Vulnerability and the Strait of Hormuz

All of which begs the question, just how vulnerable is the Strait of Hormuz? Talk of sinking ships to block the waterway is probably fanciful.

At its narrowest point, the waterway is 55 km [kilometers] wide. But the two deep-water channels for incoming and outgoing vessels are much slimmer, each 3.2km wide with a 3.2-km median between them.

While this makes for an easier target, military analysts say the Iranians would need to sink several large supertankers to effectively block the channels, an extremely difficult task.

The maximum depth of the strait is about 90 metres, while most oil tankers have a beam [maximum width] of 20–40 metres, and only a handful in the world exceed 400 metres in length.

There is no doubt, however, that Iran could make the strait unnavigable, at least for a short period.

[Since early summer 2007] Tehran's naval forces have conducted repeated exercises simulating the takeover of the strait, which is patrolled by the US Fifth Fleet and allied warships from a dozen nations deployed in the Gulf and the northern Arabian Sea.

In doing so, they have given us some idea of what tactics they might employ.

Iran has large numbers of Chinese-made C-801 and C-802 anti-ship missiles deployed in coastal batteries along the eastern shore of the waterway, aboard warships and on islands in the strait.

These would likely play a key role in any effort to block, or control, the waterway.

The narrowness of the shipping lanes makes it an ideal arena to use anti-ship missiles, because naval or civilian vessels would have little room for evasive action.

The missiles' short range minimises the prospect of US or coalition warships being able to shoot them down before they reach their targets.

Then there is the Iranian navy. Over the last year or so, coalition naval forces in the Gulf and the Arabian Sea have conducted a series of exercises designed to counter possible Iranian attempts to close the strait.

These include attacks by large swarms of small, high-speed armed craft or maritime suicide attacks similar to the Al Qaeda operations that crippled the USS *Cole*, a missile destroyer, in Aden harbour in October 2000 and blew a hole in the French supertanker *Limburg* a year later, as it sailed off the Hadramaut coast [part of the Arabian peninsula on the Gulf of Aden].

In recent months, small swarms of Iranian craft have harassed US warships in the waterway at least twice in what could have been practice runs for such hit-and-run operations, a component of Iran's strategy of "asymmetric warfare" against the superior coalition forces.

Iran claimed last October [2007] that it had amassed a fleet of 1,000 low-tech speedboats to counter the Fifth Fleet's armada of 30–40 high-tech warships.

The damage such vessels could inflict using rockets and machine guns is limited.

Broadsides of cruise missiles would be more dangerous. Iran has three frigates and 20 fast-attack craft—French-built Kaman-class vessels and Chinese-supplied Huodong boats—capable of mounting such attacks.

These vessels would be highly vulnerable to advanced US naval and airborne weapons systems if they mounted such operations.

But this does not leave much room for complacency. In the opening phase of the Hizbollah-Israeli war in 2006, the Lebanese fighters sank a Cambodian freighter and crippled an Israel missile corvette, the *Hanit*, off the Lebanese coast, using Chinese-made C-802 missiles, supplied by Iran.

Admittedly, the Israeli ship was hit because it had foolishly not engaged its anti-missile defences, but the attacks give some idea of what a larger and co-ordinated Iranian operation might achieve.

Mining the waterway is considered the most likely method Tehran would employ—a tactic that was used to some effect during the Tanker War between Iran and Iraq in their eight-year conflict in the 1980s.

The US Navy has a mine warfare flotilla based at Fifth Fleet headquarters in Bahrain, but analysts say it would takes weeks, possibly months, for them to clear safe passages in the strait if Iran was able to sow as few as 200–300 mines.

"Tehran has amassed an arsenal of naval mines and mining would be one of the most lasting and time-consuming tactics to counter," says a recent report by US security consultancy Strategic Forecasting.

"Iranian forces would use both surface and submarine assets—some more surreptitious, some less so—to attempt to saturate the Gulf."

Any such operation would send insurance rates through the roof and certainly curtail the volume of tankers using the strait if it remained navigable.

If the Strait of Hormuz were closed, only about 3m [million] barrels of oil per day could realistically be redirected by pipeline to the Red Sea.

Bypassing the Strait of Hormuz

It would also present Gulf states with the quandary of how to divert normal traffic.

If the Strait of Hormuz were closed, only about 3m barrels of oil per day could realistically be redirected by pipeline to the Red Sea.

While some transshipment traffic could move to ports such as Salalah or Aden, neither are a match for the fast-growing hubs of the Gulf coast.

"Jebel Ali has the advantage of critical mass," says Drewry's Davidson.

"In comparison with Salalah, which primarily works for one shipping line—Maersk—it has extensive services, a big market on its doorstep and even bigger markets nearby, such as Kuwait and Saudi Arabia.

"So if the ships can't get through the strait, the only real alternative is overland, and most road and rail routes are pretty undeveloped.

"This is probably why the Saudis are building new railways and new terminals to the west."

Ironically, the emergence of the Gulf as a safe harbour and transit point for international shipping owes much to the Suez crisis.

Back in the 1950s, Saudi Arabia exported most of its oil through the Trans-Arabian pipeline, or Tapline, that ran through Jordan, Syria and Lebanon to Sidon on the Mediterranean coast.

While the pipeline was not directly affected by the fighting, Riyadh [Saudi Arabia's capital] felt it was vulnerable to the Egyptian-allied regime in Syria.

It was from this point onwards that long-range oil tankers began to dominate the world energy market and the Strait of Hormuz acquired its significance.

The old route west still exists, of course. A new ... rail route linking Saudi Arabia's east and west coasts will certainly help reduce dependence on the strait for basic cargo traffic.

Similarly, the repair of old pipelines could provide Gulf producers—and their customers—some emergency relief in the event of a blockade.

But these are not ideal measures. These days, the economies of the Gulf look east to the emerging markets of Asia, and the Strait of Hormuz is their primary gateway.

More importantly, Iran is one of those economies.

[Iran] depends on the waterway not only to export its oil, but to import the refined products it is unable to produce itself.

Denied those vital fuel imports, Iranian society would grind to a halt. So while Iran could scupper [wreck] the world economy with a few well-placed mines, it would be scuttling [sinking] its own ship in the process.

11

Natural Disasters and Oil Supply Failure Could Cause a Global Crisis

The Economist

The Economist is a weekly magazine focusing on international politics and business news and opinion.

A natural disaster, such as Hurricane Katrina, that damages the energy infrastructure has the potential to send the world's economy into a tailspin. Oil supply shortages from such disasters lead to price increases, forcing consumers to cut spending which could then result in a recession. Energy experts and politicians fear the energy supply being hit by two simultaneous mishaps, such as an act of terrorism and a natural disaster.

Hurricane Katrina could lead to "one of the biggest energy shocks since the 1970s, perhaps even the biggest." So argues Daniel Yergin, a historian of oil and head of CERA [Cambridge Energy Research Asociates], an energy consultancy. He is not alone. Various oil veterans, some of whom were fairly relaxed during the demand-fired rise in oil prices from $10 a barrel in 1999 to $70 recently, are now concerned that this storm could push the oil market into a global crisis and perhaps drag the world economy into recession.

At first blush, it is hard to take such gloomy views seriously. America's economy is growing at a healthy clip, as is the world economy. What is more, history shows clearly that natural disasters like hurricanes do not have an enduring impact on the economies of rich countries. The damage done to economic output in the short term is always made up (or more) later on by a reconstruction boom. The behaviour of oil markets also called the gloomy view into question. By midweek, crude oil prices—which had risen at first to a nominal record of $70.85 a barrel—had fallen back sharply, to around $65.

So does that mean the oil sector has been spared Katrina's wrath? Not necessarily. First, Mr Yergin observes that no hurricane has ever destroyed as much of the entire value-chain of energy infrastructure, from offshore rigs to underwater pipelines to refineries and power lines on shore. On current evidence, Katrina's shockwaves will be wider, and last longer, than those of previous storms.

The world crude-oil market will not suffer particularly. Although perhaps half of the output of the Gulf of Mexico, itself accounting for about a quarter of America's oil and gas output, is still shut down, prices have not shot up, because world markets were over-supplied with crude when the storm hit. The crunch is in refining and refined products such as petrol [gasoline]. OECD [Organization for Economic Cooperation and Development] countries have agreed to release government crude-oil stocks, but America has no strategic stocks of petrol and Europe has too little to make a difference.

Katrina's shockwaves will be wider, and last longer, than those of previous storms.

Refined Oil

The world's refining sector was already stretched to the limit before Katrina hit. The problem was most acute in America, where demand has far outpaced refining capacity. As a result, America imports about a tenth of its petrol, up from just 4%

a decade or so ago. Domestic refining volumes have gone up, but not enough; and Katrina has now knocked out eight refineries, representing perhaps a tenth of the country's capacity.

Half of those refineries might be back on line within weeks—but officials say the others may take a couple of months. This would be "a huge problem, since there is not enough spare refining capacity globally," says John Paisie of PFC Energy, a consultancy. Traders are already starting to fill the gap with petrol from European and South American markets, raising prices across the world.

Fadel Gheit of Oppenheimer, a fund-management firm, is concerned. Most of the refinery workers, he points out, are still unable to return to their jobs as most lost their homes and many lost family members. He points out that the hurricane season is not over yet, and when it passes another menace will arrive: peak winter demand, which will replace demand for petrol with clamour for heating oil.

Supply and Demand

The second great unknown is how consumers will respond to short-term price rises and sporadic shortages. Already, prices in some areas have jumped a dollar since the storm, touching $4 a gallon or higher. In other places, brief shortages have been reported. With most commodities, higher prices dampen demand. Yet demand for petrol is pretty inelastic [unvarying] in the short term. Even if prices rise, soccer moms still have to go to work and drop off their children at school. Philip Verleger, an economist affiliated with the Institute for International Economics, a think-tank, suggests that it would take a doubling of petrol prices to reduce American petrol consumption by just 5%. He worries that "rising prices are going to require consumers to double expenditures on gasoline, electricity, and heating fuels. Other spending will be cut—and recession will follow."

That remains a minority view among economists; but, for the first time since the 1970s oil shocks, there was semi-serious talk in official Washington this week [September 2005] of conservation measures. The full Senate Energy Committee sent a letter to [President George W.] Bush asking him among other suggestions, to encourage federal government employees to share cars. Senator Pete Domenici, the Republican head of the committee, even floated an idea that Mr Bush hates: raising vehicle fuel-economy standards, which are close to a 20-year low.

Prices would probably jump from $60 (their assumed baseline) to over $160, even without any massive al-Qaeda attack on vital Saudi oil infrastructure.

Natural Disasters and Supply Disruption

The third great unknown about oil is whether the world will see a second blow to the system before Katrina's damage is undone. This question is ... more urgent since governments have now committed themselves to releasing a chunk of their official stocks of petroleum on to the open market. Once those are depleted, the world will be much less able to cope with the all-too-common surprises (coups, civil strife or terrorist attacks) that lead to supply disruptions in oil-producing countries.

What would happen if there were two moderate blows to the world oil economy at once? America's National Commission on Energy Policy (NCEP), a group of leading energy gurus and politicians from both parties, asked that question recently. In an elaborate exercise, the group "gamed" out the likely impacts of a natural calamity and a supply disruption amounting to just 4% of the world's oil supply. Their conclusion: Prices would probably jump from $60 (their assumed baseline) to over $160, even without any massive al-Qaeda attack on vital Saudi oil infrastructure.

At the time of the first Gulf War, such a prognosis could never have been made. Back then, global demand was much lower and the world had plenty of spare production capacity. But today, as the run-up in prices before Katrina had already made clear, there is no safety net.

12

Modernizing Electricity Management Is Key to Meeting Energy Demands

Galvin Electricity Initiative

Galvin Electricity Initiative is a nonprofit Initiative that promotes the transformation of the way we generate, deliver, and use electricity. The Initiative partners with communities, universities and entrepreneurial innovators to build smart microgrid projects based on its Perfect Power System architecture, and works with industry leaders to drive policy reform that adheres to a set of Consumer Principles.

Public and private entities joined together to develop an advanced power system named "Perfect Power System." The perfect power system is a model that can help replace the obsolete electricity grid currently used. The power demands of the 21st century include eliminating power outages, incorporating renewable energy sources, and new methods to meet rising demand and reduce electric bills.

A new approach to electricity distribution and management will provide a blueprint to eliminate power outages, incorporate renewable energy sources, meet rising demand for high-quality power and lower electricity bills.

The nation's first "Perfect Power System" is an example of how government, utilities, businesses and municipalities can

Galvin Electricity Initiative, "Public/Private Partnership Creates Opportunity To Fundamentally Address Local Energy Crises," *Transmission & Distribution World*, November 25, 2008. Copyright © 2008 Penton Media, Inc. Reproduced by permission of Galvin Electricity Initiative.

collaborate in the development and implementation of advanced power systems that are required to meet rising 21st-century power demands. The project, developed for Illinois Institute of Technology (IIT) with the Galvin Electricity Initiative, is the result of a partnership among the U.S. Department of Energy (DOE), local utility Exelon/ComEd, the entrepreneurial electricity distribution developer Endurant Energy and the Chicago-based global provider of electric power systems, S&C Electric Company.

"Power outages cost Americans at least $150 billion each year, an expense that would be largely eliminated by developing smart microgrid systems, like the Perfect Power System, that put into practice the smart technologies and distributed power generation our obsolete electricity system needs to become more efficient, reliable and secure," said Robert W. Galvin, former Motorola CEO and founder of the Galvin Electricity Initiative. "This project can be replicated in any local system—at the university, business or municipality level—once policy and technical barriers to these advancements are removed, bringing energy independence and sustainability to all customers and fostering entrepreneurship and innovation in communities nationwide."

Modern Technology and the Grid

The Perfect Power System is based on a smart microgrid—a small, local, modernized version of the electricity grid that carries bulk power across the country. These microgrids focus on rapidly bringing the economic and environmental benefits of modern grid technology to consumers. They engage entrepreneurial innovators and investors to install the smart digital technology that allows the instantaneous, two-way flow of electricity and real-time pricing and demand information between utilities and consumers. This is in stark contrast to today's antiquated, electromechanical-controlled bulk power grids that effectively hold consumers prisoner behind an iron curtain electricity meter.

"What makes a microgrid distinct from other grid improvements is that it augments remote sources of power with locally generated electricity—which may come from a natural-gas burning turbine, solar panels, fuel cells or a combined heat and power generator," said Kurt Yeager, executive director of the Galvin Electricity Initiative "This decentralized system can serve a single large building, a factory, a cluster of buildings or even a municipality that has access to and some control over its own power infrastructure and allows residents and businesses to become active consumers by giving them greater control of their power use and the ability to sell electricity back to the grid."

[The] smart microgrid distribution system will be a replicable flagship model for confronting and modeling a solution to the global energy crisis.

Innovation and the Energy Crisis

DOE recognized the path-breaking nature of the IIT Perfect Power System and its potential to spur energy entrepreneurship and expand green economies locally by investing $7 million in federal funds into the project—one of nine competitively selected to improve efficiency in the nation's power grid. IIT is investing an additional $5 million in the project. Mohammad Shahidehpour, chair, Department of Electrical and Computer Engineering, IIT, serves as principal investigator.

"As an institution that specializes in innovation based on research and technology, IIT is proud to be a pioneer in electricity delivery by applying smart microgrid engineering, like the Perfect Power System, to our energy infrastructure," said IIT President Dr. John L. Anderson. "IIT's smart microgrid distribution system will be a replicable flagship model for confronting and modeling a solution to the global energy crisis."

Projections indicate that the Perfect Power model at IIT will pay for itself within five years after it is completed. For IIT, the Perfect Power System will generate significant savings—at least $20 million over 10 years. Following the short payback period, the university will make money from Perfect Power through cheaper power costs, such as grid infrastructure improvements, allowing [it] to purchase electricity based on real-time prices rather than the traditional contracted average. IIT will also be able to sell electricity back to local energy markets and to employ more efficient energy conservation efforts by integrating local power generation from clean sources, [such as] solar.

Perfect Power at IIT is the first of several microgrid systems that the Galvin Electricity Initiative will design. The initiative is currently working in New Mexico and Massachusetts as well. The initiative also supports the Illinois Smart Grid Initiative (ISGI)—a public-private collaboration of Illinois stakeholder groups examining the benefits for consumers and the economy of a modernized electricity grid. ISGI is mapping the policy path necessary to implement smart grid technology and allow all Illinois communities to embrace Perfect Power.

13

IT Companies Become Energy Efficient in face of Global Energy Crisis

T. E. Raja Simhan

T. E. Raja Simhan is a writer for Business Line, *a financial and business daily newspaper published in India.*

Technology companies are going green, not only reducing power consumption but also reducing costs. The trend is gaining momentum as customers seek out IT suppliers who are energy efficient. Some of the methods used by IT companies to become energy efficient include consolidating data centers, reducing travel by conducting meetings over networks, and helping customers dispose of electronic waste.

Quick, how much power do you save by switching off all those unrequired systems? Enough to keep the computer plus two servers running for the next two hours?

This is the kind of computation companies, particularly technology companies, are doing extensively as they seek to wring out more work from electrical/electronic systems with less use of power, and thus lower their costs of keeping systems and devices 'cool'.

And if what is good for your bills is good for the environment too, by way of reduced emissions, so much the better.

T. E. Raja Simhan, "Watt's the Good Word!" *Business Line*, September 15, 2008. Reproduced by permission.

'Green IT' means different things to different people. Companies are approaching this from many angles—from simple moves to save power and water to [such] complex issues . . . as consolidating server space. More importantly, clients have started exerting pressure on vendors to adopt green initiatives.

Businesses around the world are consuming extreme amounts of energy through use of IT, furthering today's energy crisis.

In IDC's latest Green Poll in Asia-Pacific, 81 per cent of participating organisations said the 'Green-ness' of IT technology will become increasingly important when it comes to selecting suppliers. Some 18 per cent of organisations said they already took this factor into consideration in selecting suppliers, and another 30 per cent said they were putting systems in place to start doing so in the near future, says Sumit Mukhija, national sales manager, Data Centre, Cisco India and SAARC [South Asian Association for Regional Cooperation].

According to G.V. Singh, CIO [chief information officer], Steria India, companies are receptive to the use of energy-efficient systems, including computers. Although the IT boom in India has followed the same pattern as in the West, signs point to a shift from 'performance-per-dollar' to 'performance-per-watt' as a key metric in Indian IT purchases. Businesses around the world are consuming extreme amounts of energy through use of IT, furthering today's energy crisis, according to Steven Sams, vice-president, Site and Facilities Services, IBM. In 2007, there will be $10 billion spent on data centre energy worldwide, and IDC [International Data Corporation] predicts that power and cooling [expenditure] in the data centre will grow at eight times the rate of hardware [expenditure], he said, while launching the company's 'Project Big Green' 2.0 in India recently.

Virtualisation

Indian companies are increasingly looking to 'virtualisation' to boost their power per watt. An IDC India study has predicted that the number of virtualised servers in India will double between 2007 and 2008—rising to 45 per cent from 22 per cent.

By abstracting physical IT assets into software, virtualisation allows the consolidation of servers, thereby reducing energy costs associated with powering and cooling machines by as much as 80 per cent, says Singh of Steria. Virtualisation features are being brought into almost all IT components and services now. Cisco has consolidated most of these features into the network itself. Also, these services can now be [provided to] a wide range of users, says Mukhija of Cisco.

The data centre infrastructure is central to the IT architecture, wherein all content is stored, managed and disseminated. The components that make up the data centre are application servers, database servers, and storage and security devices.

By virtualising these resources, companies can offer services such as Intranet, Internet access and contact centres to different parts of the globe from a single centralised data centre. Consolidating resources on a single location will increase the effective utilisation of the data centre and network asset, says Mukhija.

Companies are investing huge sums of money to meet their fast expanding data centre requirements. According to industry estimates, the number of installed servers in data centres is expected to grow to 44 million by 2010, says Pallavi Kathuria, director, Server Business Group, Microsoft India.

A server uses only 15 per cent of its full capacity. Here's how. Each operating system or application requires a dedicated server for it to be run—as a result of which, companies end up deploying multiple servers to support multiple applications. While all resources of the server may be occasionally required to run the application, for the most part, the server remains unutilised when its application is not under use but keeps guzzling precious power. As a result, companies end up footing enormous electricity bills for running these power-intensive servers, leading to [waste] of resources. This is where virtualisation steps in, explains [Kathuria].

According to [Kathuria], virtualisation enables Green IT in three key ways:

- *Less [waste] and recycling costs:* Virtualisation creates multiple virtual servers on a single hardware machine, allowing multiple users to access applications from the same platform.

- *Save power:* By creating several virtual machines on one physical server, virtualisation brings down the cost of running multiple hardware machines and helps save power resources.

- *Reduced physical space:* Companies using virtualisation require less amount of physical space for storing physical machines, thereby saving on expensive real-estate investment.

'Focus on Usage'

Companies should work towards turning business 'green' through IT usage [rather] than focusing on making IT products 'green'. They should look at 'green IT' being the tool to turn their core business (services or products) processes 'green,' says Rajdeep Sahrawat, vice-president, Nasscom [an India-based global trade body].

For example, green buildings can become smart buildings through the use of sensors to monitor lighting and heating/cooling.

Similarly, technology can be used to minimise travel [and] enable employees to work from home. IT-based analysis of a company's energy usage can allow for better decision making [about] what areas to focus on to increase energy efficiency, he says.

Here's taking a look at some energy-efficiency moves by companies.

Big Blue Goes Green

'Project Big Green' includes new products and services for IBM and its clients to reduce data centre energy consumption. The company announced in August that it will redirect $1 billion a year, worldwide, across its businesses, to mobilise the company's resources to increase the level of energy efficiency in IT. IBM will consolidate thousands of computer servers onto mainframes as part of Project Big Green. It plans to save an estimated $250 million over five years. The plan is to shrink 3,900 computer servers to about 30 of IBM's 'System z' mainframes. The new server environment will consume approximately 80 per cent less energy. The discarded 3,900 servers will be recycled by IBM Global Asset Recovery Services.

IT-based analysis of a company's energy usage can allow for better decision making [about] what areas to focus on to increase energy efficiency.

Wipro's Eco Moves

Wipro plans to consolidate its data centres across the country to reduce energy consumption, according to its chief information officer, Laxman K. Badiga. Wipro's initiatives extend from energy efficient data centres to eco-friendly product engineering designs and PC ranges.

The data centres apply server virtualisation to minimise the amount of equipment required for operations. This will also reduce hardware space. The key element of Green IT is operating data centres from remote places. 'Our target this year is to reduce power consumption by 5 per cent. Last year it was 4 per cent,' Badiga . . . said.

Wipro, also a PC manufacturer, helps customers dispose electronic waste or e-waste. It identifies suitable disposal mechanisms, manages service points across India and engages disposal agencies, says the company.

Cisco Banks on TelePresence

Cisco TelePresence enables company employees to meet in real-time across continents—over [its] network. This encourages businesses to reduce the carbon footprint generated by travelling. Cisco has so far deployed more than 185 Cisco TelePresence rooms in more than 25 countries and over 85 cities worldwide. It estimates it has saved $100 million in travel costs since deploying the technology in October 2006.

NXP [formerly Philips] Semiconductors' Aerophone chip, which integrates a number of features on a single ... chip, ensures power conservation when a device goes into sleep mode. It also has power regulators that can optimise power consumption. NXP's energy-efficient solution saves approximately 85 per cent of power and soma PCs currently use these conversion chips, claims the company.

Western Digital says its 'Green Power' hard drives yield average drive power savings of 4–5 watts over competitors' drives while providing comparable performance. The power savings by one drive equates to reducing CO_2 emission by up to 60 kilograms per drive per year—the equivalent of taking a car off the road for 14 days each year.

Organizations to Contact

The editors have compiled the following list of organizations concerned with the issues debated in this book. The descriptions are derived from materials provided by the organizations. All have publications or information available for interested readers. The list was compiled on the date of publication of the present volume; the information provided here may change. Be aware that many organizations take several weeks or longer to respond to inquiries, so allow as much time as possible.

The Association for the Study of Peak Oil & Gas (ASPO)
Klintvagen 42, SE-756 55, Uppsala
  Sweden
+46 471 00 00
e-mail: mikael.hook@fysast.uu.se
Web site: www.peakoil.net

The Association for the Study of Peak Oil & Gas is a network of scientists with an interest in determining the date and impact of the peak and decline of the world's production of oil and gas. ASPO's mission is to define and evaluate the world's endowment of oil and gas; model depletion, taking account of demand, economics, technology, and politics; and raise awareness of depletion's serious consequences for mankind. The association publishes a newsletter.

Center for Resource Solutions (CRS)
1012 Torney Ave., 2nd Floor, San Francisco, CA   94129
(415) 561-2100 • fax: (415) 561-2105
e-mail: info@resource-solutions.org
Web site: www.resource-solutions.org/index.php

Center for Resource Solutions is a national nonprofit organization working to mitigate climate change. The organization builds policies and consumer-protection mechanisms in renewable energy, greenhouse gas reductions, and energy efficiency that foster healthy and sustained growth in national and international markets. *CRS NewSolutions* is a quarterly e-mail–only newsletter with updates on activities at and around Center for Resource Solutions.

International Association for Energy Economics (IAEE)
28790 Chagrin Blvd., Suite 350, Cleveland, Ohio   44122
(216) 464-5365
e-mail: iaee@iaee.org
Web site: www.iaee.org/en

The International Association for Energy Economics was founded in 1977 as a result of the 1970s energy crisis. IAEE is a worldwide, nonprofit, professional organization headquartered in the United States, with members in more than 70 nations. It provides an interdisciplinary forum for the exchange of ideas, experiences, and issues among professionals interested in the field of energy economics. IAEE publishes three periodicals throughout the year: the *Energy Journal*, the *IAEE Newsletter*, and the *IAEE Membership Directory*.

International Atomic Energy Agency (IAEA)
P.O. Box 100, Wagramer Strasse 5, A-1400, Vienna
  Austria
(+431) 2600-0 • fax: (+431) 2600-7
Web site: www.iaea.org/index.html

The International Atomic Energy Agency works with its member states and multiple partners worldwide to promote safe, secure, and peaceful nuclear technologies. IAEA publications include the *IAEA Bulletin, Nuclear Fusion Journal*, and *Fuel Cycle and Waste Newsletter*.

International Energy Agency (IEA)
9, rue de la Fédération, 75739 Paris Cedex 15
  France
(33 1) 40 57 65 00/01 • fax: (33 1) 40 57 65 59
Web site: www.iea.org/index.asp

The International Energy Agency is an intergovernmental organization that acts as energy policy advisor to 28 member countries in their effort to ensure reliable, affordable, and clean energy for their citizens. The IEA conducts a broad program of energy research, data compilation, publications, and public dissemination of the latest energy policy analysis and recommendations on good practices. IEA publications include *Natural Gas Information 2009, Coal Information 2009, Energy Statistics of Non-OECD Countries, Energy Balances of Non-OECD Countries,* and *Electricity Information 2009.*

Organization of Arab Petroleum Exporting Countries (OAPEC)

P.O. Box 20501, Safat 13066, State of Kuwait
   Kuwait
00965 24959000 • fax: 00965 24959755
e-mail: oapec@oapecorg.org
Web site: www.oapecorg.org

Established by an agreement among Arab countries that rely on the export of petroleum, the Organization of Arab Petroleum Exporting Countries is a regional intergovernmental organization concerned with the development of the petroleum industry by fostering cooperation among its members. OAPEC contributes to the effective use of the resources of member countries through sponsoring joint ventures. OAPEC publications include *Oil and Arab Cooperation, OAPEC Monthly Bulletin,* and *Energy Resources Monitor.*

Organization of the Petroleum Exporting Countries (OPEC)

Obere Donaustrasse 93, A-1020, Vienna
   Austria
(43-1) 21112-279
Web site: www.opec.org/home

The Organization of the Petroleum Exporting Countries is an intergovernmental organization. OPEC's mission is to coordinate and unify the petroleum policies of member countries and ensure the stabilization of oil markets in order to secure an efficient, economic, and regular supply of petroleum to consumers, a steady income to producers, and a fair return on capital to those investing in the petroleum industry. OPEC publications include, *World Oil Outlook 2009, OPEC Monthly Oil Market Report,* and *OPEC Bulletin.*

The United States Department of Energy (DOE)

1000 Independence Ave. SW, Washington, DC   20585
(202) 586-5000 • fax: (202) 586-4403
e-mail: The.Secretary@hq.doe.gov
Web site: www.energy.gov

The United States Department of Energy's overarching mission is to advance the national, economic, and energy security of the United States; to promote scientific and technological innovation in support of that mission; and to ensure the environmental cleanup of the national nuclear weapons complex. The DOE's strategic goals to achieve the mission are designed to deliver results along five strategic themes: energy security, nuclear security, scientific discovery and innovation, environmental responsibility, and management excellence. Popular DOE publications include, *Energy Savers—Tips on Saving Energy & Money at Home, Model Year 2009 Fuel Economy Guide: EPA Fuel Economy Estimates, Selling Energy-Efficient Products to the Federal Government—March 2008,* and *Sustainable High Performance Buildings: Leading by Example.*

The World Coal Institute

5th Floor, Heddon House, 149–151 Regent Street
London   W1B 4JD
    United Kingdom
+44 (0) 20 7851 0052 • fax: +44 (0) 20 7851 0061
e-mail: info@worldcoal.org
Web site: www.worldcoal.org/home

The World Coal Institute is a global industry association comprising the major international coal producers and stakeholders. WCI and its member companies engage with governments, the scientific community, multilateral organizations, nongovernmental organizations, media, coal producers and users, and others on global issues, such as CO_2 emissions reduction and sustainable development, and on local issues including environmental and socio-economic effects from coal mining and coal use. *Ecoal* is the official newsletter of the WCI.

World Energy Council (WEC)
5th Floor - Regency House, 1-4 Warwick Street
London W1B 5LT
   United Kingdom
(+44 20) 7734 5996 • fax: (+44 20) 7734 5926
e-mail: info@worldenergy.org
Web site: www.worldenergy.org

The World Energy Council's mission is to promote the sustainable supply and use of energy for the greatest benefit of all. The work of WEC spans the energy spectrum—coal, oil, natural gas, nuclear, hydro, and new renewables—and focuses on such topical areas as market restructuring; energy efficiency; energy and the environment; financing energy systems; energy pricing and subsidies; energy poverty; ethics; benchmarking and standards; use of new technologies; and energy issues in developed, transitional, and developing countries. WEC publications include *Energy Efficiency Policies Around the World: Review and Evaluation, Energy Policy Scenarios to 2050,* and *Energy & Climate Change.*

Bibliography

Books

Jeff W. Eerkens · *The Nuclear Imperative: A Critical Look at the Approaching Energy Crisis.* Netherlands: Springer, 2006.

Toyin Falola and Ann Genova · *The Politics of the Global Oil Industry: An Introduction.* Westport, CT: Praeger Publishers, 2005.

Ian Hore-Lacy · *Nuclear Energy in the 21st Century.* London, UK: World Nuclear University Press, 2006.

Miriam Horn and Fred Krupp · *Earth: The Sequel: The Race to Reinvent Energy and Stop Global Warming.* New York, NY: W.W. Norton & Company, 2008.

Tatsu Kambara and Christopher Howe · *China and the Global Energy Crisis: Development and Prospects for China's Oil and Natural Gas.* Northhampton, MA: Edward Elgar Publishing, 2007.

S.L. Klein · *Power to Change the World: Alternative Energy and the Rise of the Solar City.* Charleston, SC: BookSurge Publishing, 2008.

Jason Makansi · *Lights Out: The Electricity Crisis, the Global Economy and What It Means to You.* Hoboken, NJ: John Wiley & Sons, 2007.

Susan Meredith   *Beyond Light Bulbs: Lighting the Way to Smarter Energy Management.* Austin, TX: Emerald Book Company, 2009.

Robin M. Mills   *The Myth of the Oil Crisis: Overcoming the Challenges of Depletion, Geopolitics, and Global Warming.* Westport, CT: Praeger Publishers, 2008.

Sheila Newman   *The Final Energy Crisis.* Ann Arbor, MI: Pluto Press, 2008.

Bruce Podobnik   *Global Energy Shifts: Fostering Sustainability in a Turbulent Age.* Philadelphia, PA: Temple University Press, 2006.

Nathaniel Price   *The Energy Crisis: How Do We Fuel Our Future?* London, UK: Pocket Issue, 2007.

Michael C. Ruppert and Michael McClay   *A Presidential Energy Policy.* Los Angeles, CA: New World Digital Publishing, 2009.

Mary Ann Segal   *Getting Through the Wilderness: The Fuel Crisis, Global Warming, and the Hydrogen Frontier.* Bloomington, IN: Authorhouse, 2006.

Matthew R. Simmons   *Twilight in the Desert: The Coming Saudi Oil Shock and the World Economy.* Hoboken, NJ: John Wiley & Sons, 2005.

Peter Tertzakian   *A Thousand Barrels a Second: The Coming Oil Break Point and the Challenges Facing an Energy Dependent World.* New York, NY: McGraw-Hill, 2007.

Periodicals

Dennis Behreandt   "Energy's Future," *New American,* April 4, 2005.

Mary Benoit   "An Expert Look at the Energy "Crisis," *New American,* June 26, 2006.

Tsvi Bisk   "A Realistic Energy Strategy," *Futurist,* March/April 2009.

Will Brackett   "One Year Later: Pickens Plan Impacts Energy Debate," *Fort Worth Business Press,* July 20, 2009.

Kevin Bullis   "Sun + Water = Fuel," *Technology Review,* November/December 2008.

H. Stirling Burnett   "Developing Shale Oil May Solve Our Energy Crisis," *Examiner* (Washington DC), July 29, 2009.

Alasdair Cameron   "Renewables in a Time of Crisis," *Power Engineering International,* January 2009.

John Carey   "The Biofuel Bubble," *BusinessWeek,* April 16, 2009.

Tom Carney   "The Ethanol Option," *National Catholic Reporter,* October 26, 2007.

Lawrence Drake   "A Solar Pioneer Looks Back: Will
                        History Repeat Itself?" *Plumbing &
                        Mechanical*, March 1, 2009.

Rachel Ehrenberg   "The Biofuel Future: Scientists Seek
                        Ways to Make Green Energy Pay
                        Off," *Science News*, August 1, 2009.

John Gulland and   "Choosing Renewable Energy,"
Wendy Milne   *Mother Earth News*, April/May 2008.

Tariq Iqbal Khan   "The Looming Energy Crisis,"
                        Economic Review, April 2008.

Scott S. Nyquist   "Are We Headed for Another Oil
                        Shock?" *BusinessWeek Online*, July 20,
                        2009.

Cath O'Driscoll   "Fuelling Debate," *ICIS Chemical
                        Business Weekly*, March 6, 2006.

Chris Purpura   "The Smart Utility Will Be a
                        Connected Utility," *Management
                        Quarterly*, Winter 2008.

Peter C. Reiss   "What Changes Energy
and Matthew W.   Consumption? Prices and Public
White   Pressures," *Rand Journal of
                        Economics*, Autumn 2008.

Metta Spencer   "Climate Change and the Coming
                        Energy Crisis," *Peace Magazine*,
                        January/March 2007.

Gene Wolf   "Can We Achieve 20% Wind by
                        2030?" *Transmission & Distribution
                        World*, March 1, 2009.

Mortimer B.   "Energy to Burn," *U.S. News & World*
Zuckerman   *Report*, March 18, 2007.

Index

A

Adobe Systems, 14
Agriculture, 10, 64
Al-Azzaz, 26
Al-Buraik, Khaled, 29
Al Hilal Group, 68
Al Qaeda, 72
Alaska, 61
Algeria, 65
Analog Science Fiction & Fact, 31
Anderson, John L., 83
Angola, 65
Arabian Sea, 64, 72
Arsenic, 57
Athabascan oil sands, 62
Austin, Texas, 15–16
Australia, 51
Automobile industry
    batteries and, 24
    electric vehicles, 17, 23–24, 41
    energy efficiency, 33, 79
    hybrid vehicles, 17, 23–24
    nanotechnology and, 23
    public transportation and, 24
    SUVs, 33
"Axis of evil," 60, 64

B

Badiga, Laxman K., 89
Bahrain, 68, 70, 73
Baku-Ceyhan pipeline, 65
Baldo, Marc, 42
Batteries, 24
Beglinger, Nick, 42
Beingessner, Paul, 9–12

Biofuels, 49, 57, 66
Biofuels Digest, 51
Black, Tony, 60–67
Brazil, 50
Buildings
    carbon dioxide production, 42
    energy consumption, 14, 42
    greening of office buildings,
        14–16
    zero-energy buildings, 16
Bush, George W., 60, 64, 79
Business Line, 85

C

C-802 missiles (China), 72, 73
Cadmium, 57
California, 16, 46
California Energy Commission, 16
CaliSolar, 45
Cambodia, 73
Campbell, Colin, 62
Canada, 11, 20–22, 29, 62
Canadian Wind Energy Association (CWEA), 21, 22
Car batteries, 24
Carbon dioxide, 42, 58–59
Caspian Sea, 65
Center for Global Energy Studies (CGES), 27
CERA (Cambridge Energy Research Associates), 76
Chernobyl accident, 8
China
    allies, 65
    Angola and, 65
    C-802 missiles, 72, 73

energy consumption, 12, 18, 43, 63
food prices and, 51
Indonesia and, 65
Iranian oil and, 64
Sudan and, 65
Venezuelan oil and, 64–65
Christian Science Monitor, 13
Cisco India, 86, 87
Cisco TelePresence, 90
Coal
    carbon dioxide emissions, 58–59
    electricity generation, 7, 43, 56
    greenhouse gas emissions, 43
Colorado School of Mines, 62
Committee on Radioactive Waste Management (U.K.), 57–58
Computer industry, 85–90
Concentrator PVs (CPVs), 40
Concentrator solar systems, 40, 42–44
Concordia University, 23
Corn-based ethanol, 49–53
Council of Economic Advisers, 51

D

Darfur (Sudan), 64
Davidson, Neil, 70, 74
Denial, 9–10
Denmark, 22, 23
Depression, 63
DOE (U.S. Department of Energy), 50, 82, 83
Domenici, Pete, 79
Drewry Shipping Consultants, 70, 74
Dyson, Anna, 44

E

Economies of scale, 46
The Economist, 76–80
Egypt, 68, 75
Electric vehicles, 17, 23–24, 41
Electricity
    coal and, 7, 43, 56
    consumption by buildings, 14, 42
    fossil fuels and, 7
    management, 81–84
    microgrids, 82
    nuclear power stations and, 57
    Perfect Power System, 81–84
Energy crisis
    buildings as energy consumers, 14, 42
    efficient use of energy and, 13–16
    electric automobiles and, 41
    fear blocking resolution of, 54–59
    food prices and, 49–53
    fossil fuel demand exceeding supply, 9–12
    international tensions and, 60–67
    investment in oil and, 26–30
    natural disasters and, 76–80
    nuclear power and, 31–37
    paradigm shift and, 18
    peaking of oil production, 60–67
    refining limitations, 77–78
    solar power and, 38–48
    terrorist threats and, 68–75
    war and, 64–65
    See also Fossil fuels; Renewable energy; specific types of energy production
Energy entrepreneurship, 83

Ethanol, 49–53
Exelon/ComEd, 82

F

Fatka, Jacqui, 49–53
Flat-panel photovoltaic PV systems, 39
Food prices, 49–53
Fossil fuels
  agriculture dependency on, 10
  conservation measures, 79
  cost effectiveness, 7
  costs of oil production, 15
  demand exceeding supply,
    9–12
  dependence on, 7
  electricity generation, 7
  exploration for, 28
  gas prices, 52
  international alliances and, 65
  investment in oil as need,
    26–30
  Iran-Pakistan-India gas pipeline, 65
  largest oil fields, 61–62
  methane reserves, 66
  natural disasters and, 76–80
  peaking of oil production,
    60–67
  pollution and, 7
  prices, 78
  refinery limitations, 77–78
  reserve figures as inflated, 62
  sand tar projects, 29–30, 62
  sources of, 7
  supply and demand for, 9–12,
    77–79
  See also Energy crisis
France, 68–69, 72
French Petroleum Institute, 62
Freud, Sigmund, 9–10, 12

G

Gaia hypothesis, 54
Galvin, Robert W., 82
Galvin Electricity Initiative, 81–84
Gas. See Fossil fuels
Germany, 17, 18–20, 22–23
Ghawar oil field (Saudi Arabia),
  61
Gheit, Fadel, 78
Great Britain, 68–69, See also
  United Kingdom
Green Poll, 86
Greenhouse gas emissions, 14, 18,
  43
Gulf Daily News, 68–76

H

Helium, 54
Higgins, James M., 38–48
Hizbollah-Israeli war, 73
Holograms, 45
Honda, 24
Hornung, Robert, 21
Hubbert, M. King, 61
Hurricane Katrina, 76
Husain, Syed Rashid, 26–30
Hussain, Tariq, 42
Hybrid vehicles, 17, 23–24
Hydro power, 7, 21, 36
Hydro-Quebec, 22
Hydrogen, 66

I

IBM, 44, 46, 86, 89
IBM Global Asset Recovery Services, 89
Iceland, 36

IDC (International Data Corporation), 86

IEA (International Energy Agency), 11, 28

Illinois Institute of Technology (IIT), 82–84

Illinois Smart Grid Initiative (ISGI), 84

India
  allies, 65
  Angola and, 65
  energy consumption, 12, 18, 43, 63
  food prices and, 51
  Indonesia and, 65
  Iran-Pakistan-India gas pipeline, 65
  IT industry energy efficiency and, 86–87
  Sudan and, 65
  virtualization, 87–88

Indonesia, 65

Infinite Energy Magazine, 31

Institute for International Economics, 78

Intel, 46

International Auto Show (Detroit), 17

International Energy Agency (IEA), 11, 28

Iowa State University, 52

Iran, 64, 65, 68–73

Iran-Pakistan-India gas pipeline, 65

Iranian Revolutionary Guard, 69

Iraq, 64–65, 73

Israel, 73

IT companies, 85–90

J

Jafari, Mohammed Ali, 69

Jane's Defense Weekly, 69

Jane's Intelligence Review, 71

Japan, 11

JBC Energy, 30

Jebel Ali port, 70, 74

Jensen, Michael, 44

Jordan, 75

Jum'ah, Abdullah, 29

K

Kanan, Matthew, 41

Katrina (Hurricane), 76

Kooistra, Jeffery D., 31–37

Kuwait, 70, 74

L

Law for the Priority of Renewable Energy (Germany), 19

Lazear, Ed, 51

Lead pollution, 57

Lebanon, 73, 75

LEED rating system of Green Building Council, 14

Lithium, 24

Litvinenko, Alexander, 54–55

Lovelock, James, 54–59

M

Maersk, 74

Marathon Oil Co., 29

Marsden, William, 17–25

Massachusetts, 84

McKinsey & Co., 14

Mercury, 57

Merrill Lynch, 52

Microgrids, 82
MIT (Massachusetts Institute of Technology), 41
MIT News, 41
Moore, Michelle, 13–16
Motorola, 82
Mousseau, Normand, 21–22, 23, 24
Mukhija, Sumit, 86, 87

N

Nanotechnology
    automobile industry, 23
    clothing and, 47–48
    murder application, 54
    solar power and, 39, 44–45, 47–48
Nasscom, 88
Nasser, Gamal Abdel, 68
National Commission on Energy Policy (NCEP) of U.S., 79
National Corn Growers Association, 51
Natural disasters, 76–80
Nebraska, 50
New Mexico, 84
Nickel, 24
Nocera, Daniel, 41
Nonproliferation Policy Education Centre, 70
North Sea, 61
Nuclear power
    advantages, 37
    costs, 8, 66
    energy crisis solution, 31–37
    fear of, 54–59
    nuclear accidents, 8
    nuclear waste, 36–37, 57–58, 66
    pollution and, 8

propaganda about, 57–58
solution for energy crisis, 31–37
statistics on electricity production, 57
terrorism and, 36
uranium and, 11, 37
NXP Semiconductors, 90

O

Obama, Barack, 43, 47
Oil. *See* Fossil fuels
Oman, 70
OPEC (Organization of the Petroleum Exporting Countries), 28–29
Oppenheimer, 78
Organization for Economic Cooperation and Development (OECD), 11, 28

P

Pacific Gas & Electric, 46
Paisie, John, 78
Pakistan, 65
Panama Canal, 69
Perfect Power System, 81–84
Persian Gulf, 64, 72
Plastic-based solar energy conversion systems, 44–45
Platts (company), 28
Pokolski, Henry, 70
Pollution, 7, 8
Polonium-210, 54–55
Prism Solar, 45
Project Big Green 2.0, 86, 89
Prudhoe Bay (Alaska), 61
Public transportation, 24

Q

Qatar, 70
Quebec, 20, 21, 22

R

Recession, 63, 76, 78
Red Sea, 65
Reducing US Greenhouse Gas Emissions: How Much at What Cost? (McKinsey & Co.), 14
Refined oil, 77–78
Renewable sources of energy
    advantages, 7
    corn-based ethanol, 49–53
    costs, 15, 20, 24
    disadvantages, 7–8, 34–36
    Germany and, 17, 18–20, 22
    public utility company opposition to, 22–23
    statistics on electricity production and, 57
    tax incentives, 46–47
    types, 7
    See also specific types of renewable energy
Rensselaer Polytechnic Institute, 44
Richardson, Doug, 69
Rollins College, 38
Rotterdam, 70
Royal Dutch Shell PLC, 29
Russia, 61, 65, 66, 69

S

S&C Electric Company, 82
SAARC (South Asian Association for Regional Cooperation), 86
Sahrawat, Rajdeep, 88
Samotlor oil field (Russia), 61
Sams, Steven, 86
Sand tar projects, 29–30, 62
Saudi Arabia, 26, 29, 61, 70, 74–75
Saudi Aramco, 29
Scheer, Hermann, 18–19, 22, 24–25
Science Daily, 42
Shahidehpour, Mohammad, 83
Shell Oil, 59, 62
Sidon (Lebanon), 75
Silicon, 45
Simhan, T.E. Raja, 85–90
Singapore, 70
Singh, G.V., 86
Solar power
    advantages, 7
    clothing and, 47–48
    concentrator solar systems, 40, 42–44
    costs, 15, 42
    disadvantages, 34–36
    economies of scale and, 46
    efficiency, 39
    energy storage and, 41
    flat-panel photovoltaic PV systems, 39
    holograms and, 45
    nanotechnology and, 39, 44–45, 47–48
    plastic-based solar energy conversion systems, 44–45
    silicon and, 45
    solar and fuel-cell combination, 40–42
    solar collectors, 35–36
    solar-energy conversion systems, 39–40
    tax incentives, 46–47
    thin-film solar systems, 39–40
    windows and, 42–43
Somalia, 65

Spain, 22, 23
Steria India (company), 86
Strait of Hormuz, 64, 68, 69, 70, 71, 75
Strait of Malacca, 65
Strategic Forecasting, 73
Strategy + Business, 42
Sudan, 64, 65
Suez Canal, 68
Sugarcane, 50
Supply and demand for fossil fuels, 9–12, 77–79
SUVs, 33
Syria, 75

T

Tanker War, 73
Tapline, 75
Tar sands, 29–30, 62
Terrorism, 36, 68–75
Terry, Lee, 50
Thallium, 57
Thin-film solar systems, 39–40
Tolman, Rick, 51
Toyota, 24
Trafton, Anne, 41
Trans-Arabian pipeline, 75

U

UAE (United Arab Emirates), 70
Unemployment, 63
United Kingdom, 11, 54–59
Universite de Montreal, 22
University of Copenhagen, 45
Uranium, 11, 37, 57
U.S. Department of Energy (DOE), 50, 82, 83
U.S. Geological Survey, 61

U.S. Green Building Council in Washington, 13, 14
U.S. National Renewable Energy Laboratory, 40
U.S. Navy Fifth Fleet, 72, 73
U.S. Senate Energy Committee, 79
USS *Cole*, 72

V

Valero Energy Corp., 29
Venezuela, 64–65
Verleger, Philip, 78
Virtualization, 87–88

W

Water power, 7, 21, 36
Western Digital, 90
Williamson, Sheldon, 23
Wind power
  advantages, 7, 20–23
  Canada and, 20–22
  costs, 20, 24
  criticism of, 56, 58
  disadvantages, 21, 36
  electricity production, 57
  Germany and, 17, 20, 22
  limitations, 66
  water power as backup for, 21
Wipro, 89–90
World Energy Outlook, 28
World Nuclear Association, 8
World Trade Center terrorist attacks (9/11), 36

Y

Yeager, Kurt, 83
Yemen, 65
Yergin, Daniel, 76–77